BARNSLEY

GW01513386

2 016257 03

FAMOUS REGIMENTS

The
West Yorkshire
Regiment

FAMOUS REGIMENTS

Edited by
Lt-General Sir Brian Horrocks

The West Yorkshire Regiment

(The XIVth Regiment of Foot)

by

A. J. Barker

Leo Cooper Ltd, London

16. SEP. 1974

First published in Great Britain 1974
by Leo Cooper Ltd,
196 Shaftesbury Avenue,
London WC2H PJL

Copyright © 1974 by A. J. Barker
Introduction Copyright © 1974
by Lt-General Sir Brian Horrocks

ISBN 0 85052 150 5

2 016257 03

BARNSLEY LIBRARIES

PLEASE	
CLASS	356· 11094
DATE	10-74

Printed in Great Britain by
Hazell, Watson & Viney Ltd
Aylesbury, Bucks

Contents

1	Early days	1
2	Ça Ira!	9
3	Service in the West Indies	14
4	Corunna	17
5	Waterloo, 1815	21
6	India, 1807–1825	25
7	The Crimea	30
8	The Maori Campaign	35
9	Mainly South Africa	40
10	The Auxiliary Forces	46
11	The First World War	52
12	The Second World War	58
13	The Final Years	70
	Appendix A The Regimental March	
	B The West Yorkshire Regiment (The Prince of Wales's Own)	
	C Synopsis of Service	

Illustrations

facing page

1	Private Soldier, 1742	4
2	Grenadier, 1751	4
3	The Battle of Famars	4
4	Lt-Colonel Tidy	5
5	An Officer of the 3rd Battalion	5
6	Officers of the 14th Foot, Sevastopol, 1855	5
7	2nd Battalion in New Zealand	20
8	1st Battalion Guard, Dublin, 1886	20
9	Malta, 1901	21
10	Officers and NCOs of the 1st Battalion, Cambridge, 1914	21
11	The 'Leeds Pals', Glosterdale Camp, October, 1914	36
12	Canal Bank, Ypres, 1915	36
13	Captain G. Sanders, VC, MC	37
14	The 8th Battalion after the capture of the Mont de Bligny	37
15	Lt-General Sir Charles Harington inspects a Guard of Honour, York, 1925	52
16	1st Battalion Colour Party, Trim-ul-Gherry, 1937	52
17	Lord Mountbatten talks to men of the 1st Battalion, Burma	53
18	Dologorodoc Fort, Eritrea, 1941	53

Between pages 60 and 61

19	The last Colours of the 1st Battalion
20	On anti-terrorist operations in N. Malaya
21	The Princess Royal inspects the Guard of Honour
22	Suez, November, 1956
23	*Bess*, the Regiment's 'Flagship', Port Said, 1956
24	RSM Asher
25	Imphal Barracks

Introduction

By Lt-General Sir Brian Horrocks

Having read Colonel Barker's History of the East Yorkshire Regiment in The Famous Regiments Series, I looked forward to his second, which deals with their neighbours, The West Yorkshire Regiment, and I was not disappointed.

The two, of course, are now amalgamated, and form The Prince of Wales's Own Regiment of Yorkshire.

Colonel Barker is an historian of note and, unlike some of his confreres, he is not content just to reel off accounts of different battles. He brings the Regiment alive, or if you prefer it, covers the dry bones of military operations in human flesh. Since the West Yorkshires were formed by James II in 1685, as Hales' Regiment (14th Regiment of Foot) to deal with the Monmouth Rebellion, they have had many names and were associated with both Bedfordshire and Buckinghamshire until, in 1881, they found their final home in Yorkshire.

I never cease to marvel at the discipline and courage displayed in those 'days gone by' when, by modern standards, life in the Infantry of the line could hardly have been worse. Fortunately for the West Yorkshires they have nearly always been lucky in their Commanding Officers who, in those days, literally owned the Battalions, lock, stock and barrel. Moreover, the Regimental Officers purchased their promotion and their Commissions, so that all too often excellent, experienced, but poverty-stricken, subalterns remained in that rank for many years, while rich young playboys passed over their heads.

With a mean C.O. the food was terrible, and the clothing often quite unsuited to the rigours of active service. Until I read this history I had not realized, for instance, that up to 1784 no soldier was issued with a greatcoat, though a 'watch-cape' was provided for guard duties, and then taken away when the spell of sentry-go was finished.

One of the main roles of the Infantry in those days, which is often overlooked, was the maintenance of Law and Order, involving operations of a peace-keeping nature throughout the British Empire. This usually meant long periods of service overseas, in America, India, and above all, in the dreaded West Indies, where the 14th Foot spent many years, literally melting away from yellow fever. Between 1782 and 1803 they did two spells of duty in those stinking islands, amounting to 17 years in all. In the author's words, 'Finally, in 1803, what remained of the regiment returned home, under command of a Captain; behind them they left row upon row of desolate graves . . . to remind succeeding generations what tropical service meant in the old days.'

In addition the West Yorkshire Regiment played a prominent part in several of the major campaigns when Britain was involved in maintaining the balance of power in Europe – sometimes expanding to three battalions, but during the inevitable period of retrenchment which always followed every war, shrinking again to a two or even one-battalion regiment. I have no intention of attempting to describe these many famous battles, fought all over the world, as Colonel Barker does this so well, but there are a few outstanding achievements which I must touch on briefly.

The first is Waterloo, because this is an example of the innate staunchness of the young English country lad. When Napoleon, having escaped from Elba, made his unwelcome reappearance on the European scene, the great majority of Wellington's Peninsular Army was still in North America, and not available to fight in Europe. When the 3rd Battalion of the 14th, which had been raised in 1813, joined Wellington's Army in March, 1815, fourteen of the officers and three hundred of the five hundred and forty-eight other ranks were under 20 years of age, most of them Buckinghamshire lads, and the battalion was given the nick-name of 'The Peasants'. No wonder Wellington complained that his army was made up of 'worn out old men, half-witted lads and weakling boys'. Yet, these young peasants of the 14th Foot occupied the right of the line, behind Hougoumont Farm, and, in spite of being subjected all day to heavy artillery fire from the French guns, they beat off attack after attack from the French Cavalry trying to turn Wellington's flank.

When the battle was won, their Divisional Commander, General Colville, reported 'The very young 3rd Btn. 14th, in this its first trial displayed a steadiness and gallantry becoming veteran troops.'

History records that a sycophantic staff officer once said to the famous German Commander Von Moltke after he had crushed the French in a lightning campaign, 'You must be, sir, the greatest General in history.' 'No,' replied Von Moltke, 'Because I have never taken part in a retreat – the most difficult operation of all.' It is to their eternal credit that the 14th Foot have distinguished themselves, particularly in two of these very difficult operations, both fighting withdrawals in face of superior enemy forces.

In the early days of 1809 under Sir John Moore, the 2nd Battalion marched back across 200 miles of snow-covered mountainous countryside, to Corunna, constantly harassed by superior French forces. In many units discipline broke down completely, and there were thousands of stragglers. In fact the bulk of the Army arrived back at Corunna as a disorganized mob. The morale of the 14th, however, never faltered, and they played a decisive role in the final battle, fought to enable the British Army to embark, when their vigorous counter-attack saved the day. It is easy enough to be gallant during a victorious advance, but only an exceptional regiment could have survived one of the worst disasters which has ever befallen British arms, with their discipline intact.

The second occurred during the Second World War. In February, 1942, the 1st Battalion, acting as rearguard to the hard-pressed 17th Division, fought their way back for 800 miles from Rangoon to the Chindwin River. This was no ordinary rearguard action; in addition to holding up superior Japanese forces pursuing them from the rear, they had to attack and overrun a series of Japanese road blocks which the enemy had established along their line of retreat. The battalion suffered heavy casualties, and earned high praise for dogged endurance from that greatest of all Infantry Commanders, Bill Slim.

Meanwhile, the 2nd Battalion, after service in the Western Desert, arrived in the Arakan, and in the Battle of the "Admin. Box" halted the Japanese advance upon India. Both the 1st and 2nd Battalions then played a distinguished role in the defence of Imphal, which, with the supremely gallant performance of the Royal West Kents at Kohima, finally smashed the Japanese invasion of India.

No wonder the Regimental Anniversary is celebrated on June 22, when the Siege of Imphal was raised: by a curious coincidence, this was the day and month when the Regiment was raised in 1685.

Throughout history, Yorkshiremen have always been noted for their doggedness, and this quality emerges time and again throughout the History of this Regiment. Much of their success stems from the fact that, by the beginning of the 1914–18 War, their roots were deeply embedded in Yorkshire soil. No fewer than 38 battalions saw service in Flanders, France, Gallipoli and Egypt.

Unfortunately, in the 2nd World War, I never had the privilege of meeting any of the eight battalions which they raised.

I would like to conclude by repeating the opening remarks of my Introduction to the History of their present bed-fellows from the East Riding:

"This is a story of which Yorkshire can be justly proud."

CHAPTER 1

Early Days

THE story of The Prince of Wales's Own West Yorkshire Regiment begins in the year 1685, when James II succeeded his brother Charles, and the rebellion of the Duke of Monmouth gave the new king an excuse for increasing the size of his army. Ten new regiments of Foot were raised in the space of four years and one of these, first known as 'Hales's Regiment', after the name of its leader Sir Edward Hales, took precedence in 1694 as the Fourteenth Regiment of Foot subsequently bore county titles of Bedfordshire, Buckinghamshire, and finally of West Yorkshire until amalgamation with the East Yorkshire Regiment on 25 April, 1958.

James's reign was short. By 1688 most of his subjects had been alienated by his attempts to force the Roman Catholic faith on them and he was driven out of England by his son-in-law, William of Orange, who succeeded him as King William III. Colonel Hales remained loyal to James and following a series of hair-raising escapades got the deposed king safely to France. The new regime put a price on his head and when he was caught on a visit to England some time later, Hales was incarcerated in the Tower of London. In the circumstances he was lucky to get away with his life. Released after eighteen months captivity he rejoined King James in exile and died in Paris. His son was killed in Ireland, fighting for King James in the battle of the Boyne, so well remembered by Protestant Ulstermen.

Meantime King Billy had appointed one of his mercenary officers, William Beveridge, to replace Hales as colonel of the XIVth. Beveridge's command was terminated by a duel four years later. After-dinner discussions tended to become serious business in those days, especially after the port had circulated a number of times. There is no record of the cause of this particular dispute, but it appears that Colonel Beveridge became extremely angry with one of his captains, Van Burgh; a swearing match developed and then both officers drew their swords to settle the matter once and for all. The settlement ended in the colonel's death, and Lieutenant-Colonel Tidcomb from the 13th Foot took over.*

At that time and for nearly another two hundred years the British infantry was organized into regiments of foot, each of one battalion. (It was not until 1804, when a French invasion of England seemed imminent that the XIVth first raised a second battalion.) Each regiment was raised by its colonel, who regarded it as his own property and was responsible for its initial outfitting. This system inevitably meant that he had to be a wealthy man and the colonel recouped himself by selling commissions. Dress was more or less standard throughout the army, but its quality tended to reflect the whims and wealth of the regimental colonel. Like other regiments of foot – with the exception of the Earl of Bath's whose men wore blue coats – Hales's regiment probably wore collarless red frock coats, red breeches, stockings, shoes and slouch hats.

Regimental headquarters consisted of three officers of senior rank: first came the colonel, but as he was invariably an experienced soldier it was customary to employ him on responsible duties which took him away from the regiment. In practice, therefore, command in battle often devolved on one of the other two field officers – either the lieutenant-colonel or the major – to look after the day to day running of the battalion. The rest of the headquarters comprised the adjutant, the chaplain (the most frequent absentee), the surgeon and his mate, the quartermaster and the drum major. In 1696 the XIVth had thirteen companies – including a grenadier company which went in front during a march. Each company was commanded by a captain, but frequently the staff – that is the colonel, the lieutenant-colonel, the major and the adjutant – each held a company; all of them received ten shillings a day as command pay. In addition to its commander, every company had two other officers – a lieutenant and an ensign, while the grenadier company had two lieu-

---

* Van Burgh does not seem to have been indicted for murder or manslaughter. At any rate his career seems to have been unaffected, as his name appears in the casualty lists of the battle of Landen in 1693.

tenants. Roughly one quarter of these lieutenants commanded companies allotted to the staff, and did not draw command pay. Every company had an establishment of five non-commissioned officers, two sergeants and three corporals, and 50 to 60 privates. In theory the full strength of a line regiment was 44 officers and 780 NCOs and men. While it was exceptional for any regiment to be fully manned, the strength of Colonel Tidcomb's XIVth Regiment stationed at the Tower of London in 1696 is recorded as 44 officers, 104 NCOs, 780 drummers and privates and 69 servants, and its cost to the taxpayer at that time was estimated as £16,145:3:4d.

Because officers bought their commissions from the colonel they became shareholders in what was virtually a commercial venture. This purchase-system as it was called, tended to ensure a regular flow of promotion, but there were many drawbacks to it. Captains had to keep their companies up to establishment, and as recruits often cost more than the bounty money granted by the government the balance had to be made up from the captain's own pocket. Desertion was common, with men leaving one regiment to enlist in another for the sake of the bounty. As recruits were usually in demand questions were rarely asked about their past.* Another and more human reason for some of the shortfall in numbers were the one or two vacancies always reserved in each company for 'widow's men' – pay being drawn officially for soldiers who had been killed, in order to support the regimental dependants.

The men enlisted for life, and to save officers the expense of replacing those past work they were usually kept in the ranks as long as they could stand. Out of their 8d a day they paid, by way of compulsory stoppages, for food, clothing and the right to an eventual haven in the Chelsea Hospital old soldiers' home. By now the government had laid down that clothing should be provided on a two year cycle. In the first year an infantryman would be given: 'One good full-bodied coat well lined, a pair of good kersey breeches, a pair of good strong stockings, a pair of good strong shoes, two good shirts and neckcloths and a good strong hat well laced.' In the second

---

* In 1708 the recruiting problem was so acute that the government authorized the forcible enlisting of jail-birds and vagabonds.

year the more expendable items such as the breeches, the stockings and the shoes, as well as one shirt were intended to be replaced from regimental resources. With the amount of marching that was done, stockings and shoes wore out very quickly and no doubt one of Colonel Tidcomb's persistent headaches was the constant resupply of clothing for his men of the XIVth.

Dress was a matter of uniformity; so too were the weapons in current use, although the drill that went with them was a matter for individual commanding officers. During Charles II's reign two new weapons had come into use. The first of these, the bayonet, eventually resulted in firearms displacing the cumbersome pike, and by the time the XIVth embarked on its first campaign abroad, the smooth-bore muzzle-loading musket was the infantryman's principal arm. Only the name *picquet* remained behind to recall an era of close combat fighting. Meanwhile the second new weapon, the hand grenade, had led to the formation of the grenadier company in each regiment, and brought a modification in dress. As the new sort of soldiers could not throw their grenades without knocking off their broad-brimmed hats, they were given tall mitre-like caps. Only the tallest and strongest men were selected as grenadiers, and these hats accentuated their height.

In common with most of the other English line regiments, the XIVth Foot saw its first active service in Scotland where the insurgent clans supported King James. But in 1692 it sailed from Leith to join an army under the personal command of King William. His aim was to stem a French invasion of what is now Belgium, but what then constituted the Spanish Netherlands. At a time when the provision of supplies and their transport were at a premium, the Belgian terrain was popular fighting ground. The towns were dominated by fortresses and the customary method of waging war was for one side to dig in, live there at the expense of the locals and systematically lay siege to one or other of the enemy-held fortresses. If all went well during the summer months the besieged town would eventually surrender and both sides would retire for the winter. In this 'off' season the regimental officers went home to raise recruits to replace the casualties, and the following spring the same game began once more. From the officers' point of view this was a very gentlemanly arrangement. The only snag was that such wars were apt to go on for years until one side or the other was financially exhausted.

The Regiment's first notable action came at Landen in July, 1693. Despite the gallantry of the troops, King William's army was worsted in this battle. Two years later at Namur, however, the situation was reversed and the British infantry enhanced its reputation for dogged determination, and brought the XIVth its first Battle Honour.*

Namur was one of the many fortified towns in Flanders which had fallen into the hands of the French, and it was well defended when King William decided that he would capture it. Early in the siege operations 'Tidcomb's' (XIVth Foot), 'Stanley's' (XVIth Foot) and 'Collingwood's' Regiments successfully assaulted an important outpost of the town's defences. Because the assaulting troops had to cross nine hundred yards of open ground to reach their objective, the attack was launched at dusk. Grenadiers spearheaded the assault and they were followed by fatigue parties carrying fascines (bundles of brushwood) and gabions (baskets of earth). When the grenadiers had pitched their grenades over the protecting ditches and palisades of the outpost the main body of assaulting troops swarmed across the ditches filled by the fascine and gabion men and into the heart of the position. Once it was in British hands the way was clear for an attack on the citadal proper, and this was launched as soon as the fire of the batteries had breached the walls. In this final attack the Regiment was in the thick of the fighting, losing five officers and 47 men killed, and another five officers and 82 men wounded.

Two years after the fall of Namur came the short-lived Peace of Rijswijk, and when war with France broke out again in 1701 the Regiment – its numbers sadly reduced as a measure of peace-time economy – was serving in Ireland. It took no collective part in the so-called War of the Spanish Succession, but individual officers and several drafts of men were sent to bolster other regiments on active

---

* The Army Order announcing the award of 'Namur 1695' was in fact not published until February, 1910 – 215 years after the storming of the fortress.

W.Y.R.—2

service in continental Europe. In 1715 the Regiment moved to Scotland to help suppress the abortive Jacobite rebellion.

But the next outstanding action in which the XIVth played a major role was in 1727. From 22 February to 23 June that year the garrison of Gibraltar withstood a siege, whose story is a typical example of the manner in which the warnings of the 'man on the spot' are almost invariably disregarded by the authorities at home. As on many other occasions since then our soldiers were expected to make bricks without straw and to undergo perils which, with a little forethought, could have been avoided. Despite a stream of warnings about Spanish intentions from the governor, Lord Portmore, no attempt was made to reinforce the garrison or to bolster Gibraltar's defences until the siege actually began. Nine weeks later a convoy of the fleet ran the blockade of the Rock to bring six infantry regiments to reinforce the hard-pressed garrison; Colonel Clayton's XIVth Foot was among these welcome reinforcements.* A week later the Spanish opened an Alamein-type bombardment with guns which were far superior in range and accuracy to any of the British artillery. Every hour, for fourteen days, seven hundred shots were hurled into the fortress by a hundred and sixty-four Spanish guns and mortars concentrated round the defensive perimeter. Only when these weapons began to break up under the stress of maintaining this fantastic rate of fire did the bombardment begin to slacken. In that fortnight diplomacy had also been at work and by the end of July a truce had been agreed and the safety of the fortress was assured. But the garrison had had a hard time and before the Spanish lifted the siege, the Regiment had suffered 28 casualties.

For the next fifteen years the Regiment continued to garrison the Rock, returning to England in 1742 to be quartered in and around York. Three years later, Colonel Joseph Price, who had succeeded General Clayton in 1743, was ordered to take the XIVth to Flanders where it joined the Duke of Cumberland's army. This was a brief

---

* Jasper Clayton, a distinguished officer and good soldier, succeeded Tidcomb on the latter's demise in 1713. Subsequently Clayton became Governor of Gibraltar, being responsible for its defence in the siege of 1727. As a Lieutenant-General he was killed at Dettingen.

interlude. When the Duke was defeated at Fontenoy, his campaign ground to a halt and the bulk of the British troops were hurried back to Scotland to meet Prince Charles Stuart's invasion. In this petty rebellion the Regiment took part in the decisive battles of Falkirk and Culloden.

Meantime a royal warrant, published in 1743, had brought important changes in dress and rationalized the use of regimental Colours. In the early days, individual infantry companies customarily carried two flags – that of their colonel and that of the company commander. Gradually this habit had fallen into disuse and the number of flags was reduced to two per regiment. Under the new warrant the Colonel's flags were abolished and the use of armorial bearings forbidden. In their stead regiments were authorized to carry two Colours: the 'King's', based on the 'Union' flag, and the 'Regimental', embroidered with the regimental number.

From Scotland in 1752, the XIVth marched south – a brave sight in their scarlet coats with buff cuffs, scarlet breeches, tricorn hats, buff belts, buff pouches and white gaiters. In accordance with the latest royal warrant the grenadiers wore 'the King's cipher and crown' in front; the Hanoverian White Horse with the motto '*Nec aspera terrent*' on the flap and the number '14' behind.

Back in England the Government had use for idle soldiers and the Regiment was set to work helping Customs officers to suppress the smuggling, then flourishing on the south coast and depriving the Inland Revenue of taxes. More garrison duty in Gibraltar preceded a tour in the West Indies, – remote and unhealthy colonies where there were no alehouses and where barrack accomodation was makeshift. On St Vincent the XIVth fought a campaign against the Caribs, but their most exacting enemy was disease. They had to eat the food of the country, some of it quite unsuited to British stomachs. Yellow fever and dysentery were rife, and the ravages of 'Yellow Jack' produced more casualties than the Caribs. In many cases once a regiment was sent to the West Indies it stayed there for generations. Fortunately for the Regiment, however, the American War of Independence brought a move to a more salubrious climate. The XIVth took no part in any of the major actions which ultimately decided the destiny of the New World. But today, nearly two hundred years after these momentous events, one of the lesser known exploits that led to the Declaration of Independence is especially worthy of this chronicle. At dawn on 8 December, 1775, a detachment of 120 men of the XIVth attempted to clear an American position at Great Bridge, near Norfolk in Virginia. This force was under the command of Captain Charles Fordyce and Lieutenant John Batut was at the head of the column. When Batut crossed the bridge the Americans opened fire, killing Fordyce and twelve of his men within minutes of the action starting. When Batut and sixteen other soldiers were wounded and captured, the rest of Fordyce's men fell back across the bridge to a position defended by Captain Samuel Leslie. That was the end of a scene which has long been relegated to the limbo. As a sequel it is worthy to recall that eighteenth century military eitquette allowed Fordyce and the other casualties a burial 'with full military honours'.

In 1776 the British troops in North America concentrated at New York and when the time came to pull out of the New World the XIVth was among the first regiments to leave. Decimated by service in St Vincent, most of the remaining fit men were drafted to other units before the Regiment sailed from North America, and only a cadre returned to England to recruit and reform.

CHAPTER 2

Ça Ira!

THE final convulsion of the war-racked eighteenth century was the French Revolution, and once again the British Army became involved. When the Republicans invaded the Netherlands the British Government decided that British trade interests were threatened and troops were sent to support the Austrians and Dutch already fighting in Holland. The expedition, composed of British, Hanoverians and Hessians – all under the command of the Duke of York – was not particularly successful. Twenty-eight years old, His Royal Highness had studied the profession of soldiering in Berlin and was a thorough partisan of the red tape and pipeclay system of the Prussian army. Powdered hair and pigtails were the order of the day; grenadiers wore side whiskers; smartness was the criterion, and little thought was given to comfort. The fact that no soldier in those days possessed such a thing as a greatcoat is worthy of record. The nearest approach was a 'watch-cape', issued for guard duty and taken away when the spell of sentry-go was finished. In the Dutch winter of 1794 deaths from exposure and cold were so numerous that the Duke of York harried the Government into a general issue of greatcoats – pointing out that the expense would be more than repaid by the lives saved.

Disembarking on the island of Voorn in April, 1793, the XIVth Foot was the first British line regiment to take the field. And according to 22-year old Lieutenant Thomas Powell the locals appeared to be more sympathetic towards the French than to Colonel Hotham's men. The Regiment's first action came with the attack and capture of the French entrenched camp at Famars. Famars was close to the border town of Valenciennes, the capture of which Prince Frederick, the Allied commander, regarded as the preliminary step to a triumphal march on Paris. The garrison of Famars were Revolutionaries –

ragged, starving, fanatical, desperate – with little to sustain them but the spirit of hate and the music of their bands.

One of the favourite war-songs of that period was the famous *Ca Ira* – the Tipperary of its time. At first a mere chant extolling liberty, it had swung into the popularity chart of 1790 when a comparatively innocuous ditty was sung by French workmen during the 14 July celebrations to the dance tune *Carillon*. Three years later the words to the refrain had a savage and anarchistic message: '... The aristocrats to the lamp-post ... We will hang them ... if not hanged, smashed; if not smashed, they will be burnt'. This was the anthem that was sung by the garrison of Famars and the spirit engendered by it all but overwhelmed the XIVth.

The attack on Famars was launched at dawn in a misty half-light and the initial assault was completely successful – seven cannon and 200 prisoners being taken. But the French regrouped and counter-attacked with dervish-like frenzy and for a moment the Regiment wavered. Then the CO, Lieutenant-Colonel Charles Doyle took over. Doyle had decided that *Ca Ira*, the song of victory, must be matched against itself: 'Come on, my lads', he shouted, 'let's break the scoundrels to their own damned tune. Drummers, strike up *Ca Ira*'. The drummers obeyed; the effect was irresistible and the counter attack was defeated. Famars fell and, by express order of the Duke of York, the air was adopted as the Regimental quickstep. As such it is played by the amalgamated Regiment to this day.

From Famars the British expeditionary force moved up to besiege Valenciennes and the XIVth was ordered to provide one of the assaulting columns for an attack on the outer defences. Doyle, who was clearly something of a fire-eater, elected to command this force himself, and at a parade of the whole Regiment he called for a hundred volunteers 'for a service of the greatest danger'. After explaining that the objective was mined and that the attackers could expect to be blown up when the French fell back, he ordered those who were prepared to volunteer to 'Recover Arms'. Every man promptly brought his musket to the 'recover', and Doyle was said to have been considerably moved by this display of enthusiasm. A hundred men were selected and the attack was launched. In the event the French had no time to blow their mines and the losses incurred in the attack were negligible.

After Valenciennes the original plan started to go awry. Before there could be any march on Paris it was decided that Dunkirk would have to be taken. So the British Army marched south to Cambrai and then turned back to the coast. In a forced march of this nature 40 miles a day was not uncommon, and in his diary Lieutenant Powell has left a brief record of the hardships involved. The roads of 1793 cannot, of course, be compared to those of modern times. Engineers and pioneers had to precede the column to patch up the bridges and to fill in the worst of the pot-holes to smooth the way for the wheeled transport. Because of the atrocious state of the roads there was generally no question of moving after dark. In any case the men were too tired. The hardest punishment was meted out for minor breaches of discipline; nevertheless this was a perennial problem. Opportunities to supplement the rations were a great temptation, and Powell records seeing the men of a Hessian regiment sneaking off to 'liberate' a flock of sheep. In the operation which followed, one subaltern, Lieutenant Tom Clapham, is reported to have jumped into a water-filled ditch, so that the grenadiers could use his back as a stepping-stone. This quick witted action probably contributed to the speed of the advance.

An attempt was made to take Dunkirk, but when a determined attack had been repulsed with heavy loss the siege was lifted. (The Regiment suffered fifty-six casualties in this action.) The army now retraced its steps, marching to Oudenarde which, so far as the XIVth was concerned, became a staging post for its next action at Tournai.

'Tournai' is the Regiment's second Battle Honour, and it was earned in a week of fighting round Lille. On 18 May, 1794, four days before the actual battle for Tournai, the allied army found itself outnumbered and out-manoeuvred. The Regiment, together with the 37th Foot (The Hampshires) and the 53rd (The Shropshire Light Infantry) was part of Major-General Henry Fox's 4th Infantry Brigade at this time. This brigade was advancing for a concerted allied attack on the French positions. But the timing of the allied plan went wrong and Fox's brigade found itself isolated in difficult swampy terrain, criss-crossed with rivers and drains. Fox decided to turn back and the brigade began an orderly withdrawal. But the French had cut the road down which the troops were retreating and the XIVth were stopped by heavy fire from positions covering a formidable road block. The brigadier, who had ridden up to see what was holding the withdrawal, appears to have been unduly pessimistic. 'I fear we shall have to lay down our arms' he said to Captain Tom Clapham, now commanding the advance guard. 'No, Sir!' the ubiquitous Clapham replied, 'The Fourteenth can go through them.' And go through them they did. Led by Corporal Gilbert Cimitière, a French emigrant serving with the XIVth's grenadier company, the Regiment left the main road and circled round the French position. One hundred and fifty years later another generation of the XIVth were to perfect this technique of encirclement in Burma. In the event Clapham's initiative paid off and, although the move was not accomplished without loss, the brigade eventually slipped out of the French net. Corporal Cimitière was awarded a commission for his part in the action and eventually rose to command the 48th Regiment.

On 19 May, the Allied force was concentrated in the immediate neighbourhood of Tournai, and the French were attempting to press home their success of the day before. In the early morning of the 22nd, they attacked in four columns, and inevitably the French numerical superiority began to tell. Gradually the Allies were forced back. In a desperate effort to regain the initiative four allied brigades were ordered to recapture the village of Pont-à-Chin, overlooking the valley of the Scheldt. As a result of their fight to break out of the French trap on the 18th, the 4th Brigade was down to about 600 men and up to this time had been held in reserve. But it too was now thrown into the fight, and the three regiments concerned raced forward to capture the village at the point of the bayonet. In the fracas one man, a certain Private Tovey, took time off to knock down a hare which ran out in front of him and stuff it in his haversack. Meanwhile the villagers, it is recorded, watched 'in painful suspense' as the troops swept past them to storm the French battery located near a windmill on the hill above their houses. The battle cost the Regiment 5 killed and 30 wounded that day, but it undoubtedly brought a change in the allies' luck, for by nightfall the French had been beaten back with a loss of 6000 men. So ended Tournai. 'But for this handful of British soldiers', wrote Sir John Fortescue, 'the day would have been lost to the Allied. But whether this be true or not, 27 May, 1794, should be a great anniversary for the Fourteenth.'

CHAPTER 3

The West Indies

So far as Britain's Prime Minister was concerned the campaign in Europe during the War of the French Revolution was of secondary importance. He believed that a country in such financial difficulties as France could not maintain a war for long and that the best way of reducing her to submission was to capture the French islands in the West Indies – so destroying France's colonial trade. This task promised to be made easier by the fact that the French Revolutionary Government had proclaimed the equality of black men with white, with the result that the blacks had risen against the whites and the whites were prepared to welcome anyone who would restore their dominance. Unfortunately nobody in Westminster seemed to appreciate the fact that two years of the West Indian climate was enough to destroy a battalion, even in time of peace.

In 1793 seven thousand British troops and a fleet had been sent to occupy the French West Indies. By the end of 1794 five thousand of these men had died and a year later the British had been dispossessed of most of their own colonies in that part of the world. Dribbles of reinforcements only helped to fill the graveyards and eventually a fresh army of seventeen thousand men was sent out to recover the lost ground.

The Regiment, which had returned to England in 1795, was part of this force. Owing to a series of blunders – one of which nearly resulted in the loss of the ship carrying the XIVth – the expedition started late and it was April, 1796, before operations got under way. Sir Ralph Abercromby, an officer of considerable experience who had commanded the 4th Brigade in the Duke of York's army in Flanders and who held the Regiment in high regard, had been put in charge. Faced with the task of re-conquering practically the whole of the French West Indies, his enemy were not so much the troops of the French Republic but the hordes of negroes whose passions had been inflamed by revolutionary doctrines. In the event, Abercromby's first operation was directed against St Lucia, where a small assault group from the XIVth and 28th Regiments was put ashore on 26 April. In the course of a week the invaders had been built up to a force of 18,000 men under command of a dynamic young brigadier whose name is now more usually associated with the Peninsular War. Following an engagement with the French shortly after disembarkation, the Regiment saw little action and their casualties were limited to five killed and two wounded. During the ensuing year yellow fever was to prove far more deadly.*

Despite the steady erosion of the army's strength, the West Indies campaign continued. In August, 1796, Spain had ranged herself alongside France and the following year Abercromby was ordered to capture Trinidad. Meanwhile on Puerto Rico the Regiment was involved in a most unusual action. Abercromby had established a fortified base on the island, whose garrison were preparing for the next phase of the operation when a Spanish bombardment created havoc in one of the British outposts. Two guns were dismounted and the protective earthworks pulverized. To repair the damage a working party of 150 men was supplied by the Regiment. Under cover of darkness they set about their task. Meantime some five hundred Spaniards had managed to slip up to a patch of undergrowth about a hundred yards from the position. There they remained undetected until dawn, when the picket covering the working party was called in. At this particular time the attention of everybody labouring to get the guns back in action was focussed on Abercromby, who had just arrived with a posse of staff officers to see how things were going. Consequently nobody was ready when the Spaniards broke cover and charged the position.

Only the men who had been on picket duty were armed and the Spaniards were all brandishing knives or swords. However, the working party quickly recovered and turned on their assailants with picks and shovels. At the cost of five men killed and seventeen

* Thomas Powell, now a captain, distinguished himself in this action. He, too, fell a casualty to sickness and had to sell his commission.

wounded every Spaniard was accounted for – killed, wounded or taken prisoner. Needless to say this episode served to enhance the Regiment's reputation in Abercromby's eyes, and when he was sent to Egypt to attack the French army there in 1800, he wrote 'regretting' that the XIVth were not on the expedition with him. 'I do not think any service can go on well without them,' he concluded.

It was perhaps unfortunate that Abercromby had to leave the Regiment behind when he returned to England. Puerto Rico was never captured and with the passing of time the army in the West Indies was steadily reduced to a shadow. So terrible were the losses incurred from sickness that it was with much relief that news was received of the signing of the Treaty of Amiens. The bickering for the Caribbean islands had ceased, but they still had to be garrisoned. So the XIVth soldiered on, moving first from Puerto Rico to Martinque, and from there to Trinidad. Finally, in 1803, what remained of the Regiment returned home under the command of a captain. Behind them they left row upon row of desolate graves to remind succeeding generations what tropical service meant in the old days.

CHAPTER 4

Corunna

THE truce patched up by the Treaty of Amiens was short-lived. In May, 1803, Napoleon declared war on Britain, and the so-called 'Grand Army' started to concentrate around Boulogne for an invasion of England.

To meet this threat the British Army was expanded, and at Belfast in 1804 the Regiment formed a second battalion. Volunteers were enlisted by recruiting teams sent by the 1st Battalion to towns as far afield as Leicester, Lichfield, Manchester, Northampton and Norwich. But the response was disappointing until the government introduced measures to stimulate recruiting. Up to this time a soldier joined for life; under the new regulations a man could enlist for 'short' service – seven years with the colours being considered 'short' in the eighteenth century – the only snag being that his seven years did not count towards a pension unless he re-enlisted on a normal engagement. The scheme was novel and, although it was hardly as successful as the government had hoped, it did produce some recruits and the Regiment benefitted. In 1806 the 2nd Battalion went to Ireland barely 200 strong; the effect of the new recruiting measures brought a draft of 500 men, 'fine volunteers from Bedford, Berkshire, Hertford and Notts militias'.

Meantime the 1st Battalion, which had been deployed on the south coast in anticipation of the French invasion, had been sent to Northern Germany. There it constituted part of a force whose aim was to recover Hanover from the French. Nothing came of the expedition and when the Battalion returned home it was also sent to Ireland. (As is usual in that unfortunate country, religious and political feelings were running high, and Westminster saw Ireland as Britain's Achilles heel.) But it did not stay long, sailing in June, 1807, for Madras under the command of Lieutenant-Colonel James Watson 'with an embarkation strength of forty-nine officers, sixty-four sergeants, twenty-two corporals, fifty-nine drummers and one thousand and fifty-five privates, all "life-service" soldiers'.

In India the 1st Battalion took part in an attack on the Danish settlement of Tranquebar in the Carnatic. It was an unspectacular affair and, once it was over, the 1st Battalion spent the next eighteen months on garrison duty in Bengal. As the news of Moore's campaign in the Iberian Peninsula slowly made its way to India, the men of this battalion may well have been depressed when they realized they were stationed in a part of the world where there was little prospect of further active service. Yet their lot was common to the greater part of the British army scattered over the face of the globe.

The 2nd Battalion was more fortunate. In October, 1808, it embarked at Cork and landed at Corunna in Spain to reinforce the British Army in the Peninsula then under command of Sir John Moore. Napoleon had invaded Spain in February, occupied Madrid, proclaimed his brother King of Spain, and gone on to capture Lisbon in Portugal. When the Spanish turned to Britain for assistance Wellington was sent to recapture Lisbon and drive the French out of Spain. Left to himself the Duke might well have succeeded in doing so, but at a critical juncture he was superceded in command, and Napoleon was given a breathing space. Assuming personal command, the French Emperor lost no time in regrouping his troops for a campaign to finish the war in Spain once and for all. By the time Lieutenant-Colonel Jasper Nicolls' 2nd Battalion of the XIVth arrived in Spain, Napoleon's army of 150,000 French soldiers had smashed the Spanish armies and was ready to turn on the British. Thus, when the 2nd Battalion joined Moore's army near Salamanca on 20 December, 1808, the French were advancing in overwhelming force from Madrid.

Short of supplies and transport and lacking the support he had expected from the Spanish, Moore had already decided to run for it. Retreating into Portugal was impractical because of the bad roads and likelihood of being outflanked. So he decided to retire northwards to Vigo and Corunna and get his army out of Spain before Napoleon enveloped it. Marching back across two hundred miles of the desolate snow-covered countryside, soaking wet, footsore, tired to the point of exhaustion, and hungry, the troops had a rough passage – especially when they crossed the Galician mountains. In many units discipline broke down and there were thousands of stragglers. In consequence, by the time the army trailed into Corunna on 11 January it was little more than a disorganized mob.

Moore had expected to find ships waiting to evacuate his army, but the British fleet did not arrive until the 14th. Embarkation started at once; meanwhile the few days respite had done much to restore morale and discipline, and when Marshal Soult tried to interrupt the evacuation the French were repulsed.

The French attacked in three columns during the afternoon of 15 January, 1809, and the battle lasted for four hours. The Regiment, as part of General Viscount Hill's 4th brigade (with the 5th (Northumberland Fusiliers) and the 32nd (Cornwall Light Infantry)), was holding the left of the defensive perimeter around Corunna. The main road to Madrid ran through the village of Palavia Abaxo, at the centre of the XIVth position, and it was here that one of the attacking columns concentrated its efforts. The column appeared to consist of about a battalion, formed on a depth of twelve men and a front of sixty. Crossing the brook which lay at the foot of the ridge defended by the XIVth, it slowly began to ascend the slope. Within a short distance of the crest, about a hundred yards from the village, the shakos, blue coats and white trousers of a French line regiment were clearly visible, as were the cocked hats and trailing scabbards of the officers. At this point the XIVth went into action. British muskets came up to the 'Present', there was the crash of a volley and the French column was veiled from sight in a cloud of smoke which was ripped and torn by two more discharges. When the air cleared the hostile column, momentarily checked, was seen to be regrouping to continue its advance. In the end the attack was driven off, but it was a close call, and at one time the French actually got into the village, and were only driven out by a counter-attack led by Jasper Nicolls in person. This particular action was a complete success, marred only by the untimely death of Moore, who was buried to the sound of French guns a few hours before the Regiment embarked. That night, Sir John Hope, who succeeded Moore in command, wrote in his despatch that 'It is particularly incumbent on the Lieutenant-General to notice the vigorous attack made by the 2nd Battalion 14th Foot, under Lieutenant-Colonel Jasper Nicolls.' Regimental records also mention that Colour-Sergeant Thomas Garrett of the Regiment distinguished himself in the same action.

That was the end of the first phase of the Peninsular War. The Regiment was awarded the Battle Honour 'Corunna', but because it did not return to Spain it was debarred from the distinction 'Peninsula'. A gold medal was conferred on Jasper Nicolls, and eventually, in 1847, seventy-nine survivors of Corunna received a silver medal.

Hill's brigade was the last to leave Spain, and the XIVth covered the final embarkation on 17 January, 1809. Four days later the ships reached Portsmouth and the people there were shocked by what they saw. Men and officers alike were in rags, unshorn, filthy and covered with vermin. But there was more to the public outcry that followed. The fact was that both the Army and the public were getting tired of expeditions which embarked, landed, re-embarked and returned with nothing accomplished. In this instance the criticisms that were voiced were unjust. Despite enormous odds Moore had managed to extricate 23,000 out of a total of 29,000 men from an untenable position and the troops had done everything that had been asked of them – and more. In particular the Regiment's performance outside Corunna was magnifient. It is useless disguising the fact that the French claim Corunna as a victory. But, for the XIVth, it was not so much a defeat as a great military achievement.

CHAPTER 5

Waterloo

WATERLOO is generally considered to be the greatest of all battles fought by British troops and it has been described in the minutest detail by the most accomplished military historians in Britain, Germany and France. As is well known, Napoleon, worn down by the successive campaigns which had been waged against him in Europe and Spain, had abdicated and retired to honourable exile in Elba. In the early spring of 1815 he violated the conditions of his exile, and returned to France where the majority of his soldiery flocked to his standard. Once more the Allies mobilized their armies and prepared for war.

In Britain, the Regiment had formed a third battalion in 1813, and when Napoleon re-appeared on the scene this battalion was in Plymouth waiting to be disbanded as part of the economic retrenchment which inevitably follows any war. The commanding officer was Lieutenant-Colonel Francis Skelly Tidy, who had spent the greater part of his twenty-two years of service in the West Indies. According to Lord Albemarle, who was one of his subalterns at Waterloo, Tidy was a 'good-looking man . . . with a spare but athletic figure, . . . of frank and agreeable manners'. That he was something of a character and extremely popular in the Regiment, there can be little doubt.

In March, 1815, the 3rd Battalion was ordered to join Wellington's army which was then assembling in Belgium, and in due course it arrived in Brussels. Fourteen of the officers and 300 of the 548 other ranks of the battalion were under twenty years of age. Most of them were from Buckinghamshire, fresh from the plough and called at home 'The Bucks', although their unbuckish appearance had also led to them being nicknamed the 'Peasants'. Some days before the battle the Battalion paraded in the Brussels Grande Place to be reviewed by Wellington's Inspector-General, the ageing General Mackenzie. 'I never saw such a lot of boys,' Mackenzie complained as he stumped round the ranks. Tidy asked him to tone down his criticism, 'Very well,' said the Inspector-General, addressing the troops, 'I called you boys. And so you are. But I never saw so fine a set of boys, both officers and men.' This was as far as Mackenzie was prepared to go; he was not prepared to pass the Battalion as fit for active service and Tidy was told that his unit was destined for the Antwerp garrison. This prospect had no attraction for Tidy and he appealed to General Hill. 'Were you satisfied with the behaviour of the XIVth at Corunna?', Tidy asked. Lord Hill assured him that he had been more than satisfied. 'Then,' said Tidy, 'I am sure you will save this fine regiment from the disgrace of garrison duty.' Hill agreed to do his best; Wellington himself was asked to review the Battalion, and the Duke countermanded Mackenzie's order.

The XIVth was not the only regiment filled with young, untried and barely trained troops. Wellington complained bitterly that his army was made up of worn-out old men, half-witted lads and weakling boys. The fact was that most of the experienced soldiers were in North America. And, in spite of the inducements and reforms introduced by a government anxious to meet the crisis precipitated by the advent of Napoleon, very few able-bodied men wanted to join the army. Pay was poor, living conditions in the newly built barracks were bad, and discipline was maintained by the lash. Many men tempted to join up for a life of adventure were put off because they were afraid of being sent to the West Indies and certain death by yellow fever. There were no good-conduct badges and no authorized good-conduct pay, although in the XIVth the officers devised a system which helped to make good this defect. It was only later, on Wellington's recommendation, that a higher grade of NCO – the colour-sergeant – was introduced to encourage good men.

The battle of Waterloo began at 11 am on the morning of 18 June, 1815. The 3rd Battalion, with the 23rd (Royal Welsh Fusiliers) and 51st (Kings Own Yorkshire Light Infantry) formed Colonel Harry Mitchell's 4th Brigade in General Sir Charles Colville's 4th Division. Briefly what happened was as follows: Napoleon attacked, and for four long hours the British army withstood the onslaught until the Prussians arrived. When the Prussians went into action, success was assured; by sundown the battle was won, the French were in full retreat and Napoleon's sun had set for ever. The crucial period was the four hours during which Napoleon tried to tear Wellington's army to pieces and to sweep it away by successive attacks of infantry, cavalry, mixed infantry and cavalry and finally the infantry of the Imperial Guard. Never have the British given a finer example of their power of endurance.

When the battle began the Regiment was deployed on the right of the line near the Nivelles-Hougoumont road. One company formed the right of a line of skirmishers and the remaining companies were in reserve, located to gain the protection of some dead ground. French artillery hammered the area all day, but fortunately the XIVth suffered few casualties. (Unlike the 27th Iniskilling Fusiliers who stood in square formation throughout the bombardment, 'never moved a step and never fired a shot'. Out of 700 present they lost 478 killed and wounded.) In an attempt to break through to Hougoumont, French cavalry repeatedly charged the right of the Regiment's position during the early afternoon. One party of French cuirassiers actually broke through the line but were compelled to surrender.

The struggle reached its intensity about 3 pm, and for a while the centre of Wellington's position was in danger of collapse. A redeployment was ordered and the XIVth took up a new position, forming a square in the middle of the plain 'about a 100 yards from the Nivelles road'. 'Our intentions were to look out for the Cavalry of the Imperial Guard . . . a magnificent body of horsemen, who advanced towards us at the *pas de charge*. For a moment we were in doubt as to which square they intended to honour, but they gave preference to our neighbours, a regiment of Brunswickers. After vain attempts to pierce their square they wheeled fifty paces to our rear . . . When the smoke cleared away the Imperial horsemen were seen flying in disorder . . .'

By now the critical moment had passed; the Prussians had arrived and were attacking the angle between the French right and centre. Finally, when Wellington ordered his whole line forward to the counter-stroke, Napoleon's troops wavered and dissolved into a panic-stricken mob of broken fugitives. That night the 3rd Battalion of the XIVth bivouacked just outside Hougoumont. Colonel Tidy had gone into action with 38 officers, 33 sergeants, 11 drummers and 548 rank and file. Casualties that day amounted to seven men killed, with one officer and twenty-one men wounded.* The Regiment's performance was aptly summed up in a Divisional Order published by General Colville two days after the event: '. . . the very young 3rd Battalion 14th, in this its first trial, displayed a steadiness and gallantry becoming veteran troops . . .'

Acknowledgment of this steadiness and gallantry was reflected in the award of the dignity of a newly created Third Class of Bath, better known as the CB, to Lieutenant-Colonel Tidy. Every other officer and man who was present at Waterloo also received a credit of two years service and a silver medal.

With the rest of the 4th Division, the 3rd Battalion marched gaily to Paris with the band playing *Ca Ira*. There they camped in the Bois de Boulogne and participated in the victory celebrations which followed. By Christmas 1815 it was home, and two months later it was disbanded.

* These figures are at variance with those given in previous regimental histories. They were taken from the official records of the Public Record Office.

CHAPTER 6

India

WHILE the 3rd Battalion was earning fresh laurels for the Regiment at Waterloo the 1st Battalion continued to soldier on in India. The 2nd Battalion, ordered to Malta in 1819, took part in an operation against Marseilles during July, 1815. But when Napoleon surrendered, this battalion returned to Malta. In 1817 this post-war economic purge brought disbandment, and its redundant 420 men were transferred to the 1st Battalion. Thus in December, 1817, the 1st Battalion was again the sole custodian of the Regimental prerogative.

Apart from taking part in campaigns against the French in Mauritius in 1810, and the Dutch in Java in 1811, the 1st Battalion served in India for nearly twenty-five years. And when they returned to England the fact that there was no official acknowledgment of this service rankled with at least one officer. In 1833 the Regiment was at Portsmouth, brigaded with the Queens (2nd Foot) the Kings (8th) and the Royal Marines. Colonel Everard, who commanded the battalion at that time, firmly believed that the Regiment should have been awarded the title 'Royal Light Infantry'. At a garrison parade when Commanding Officers were addressing their battalions as 'Queen's', 'King's' and 'Royal Marines', Everard shouted 'Neither "King's" nor "Queen's", nor "Royal Marines" – only plain XIVth that have done twenty-five years in India and never unfixed bayonets – "Attention".'

Apart from the non-recognition of an extraordinary long and arduous tour of foreign service, several campaigns were ignored. There was no award for the capture of Mauritius, for instance. But the subjugation of Java, which cost the Regiment twelve killed and ninety-five wounded, did in due course bring the Battle Honour 'Java'. In effect this was earned at the great entrenched camp of Cornelis, where French and Dutch had prepared to fight to the last.

The fighting there went on for eleven days, during which as many men of the XIVth succumbed to the heat and sheer fatigue as were pronounced battle casualties. In the final assault the Regiment incurred sixty-four when a captured ammunition dump was blown up, and for his part in the action Lieutenant-Colonel James Watson, the CO was commended and received a gold medal.

While this battle virtually broke the Dutch in Java, it was some time before the country was pacified. In 1812 the Sultan of Mataran, a provincial potentate, decided that he would drive all Europeans off the island. Three miles of fortifications and a great ditch enclosed his palace, 'Crattan', and the Sultan had assembled 17,000 well-equipped troops and a hundred cannon to defend it. He could also rely on the support of the local population, many of whose 100,000 men were armed. To counter this threat the British could muster a mere 1,500 men of the XIVth and 78th Regiments. But daring paid dividends. Under James Watson's command a composite column from the two regiments attacked the Crattan, and after a good deal of hard fighting captured both palace and Sultan. That was the end of the Sultan's ambitions, and the Battalion returned to India the following year.

Trouble was always brewing in the Indian sub-continent, but as far as the Regiment was concerned the next involvement came in 1814, in a war against Nepal. For some time before hostilities actually erupted, the Gurkhas had been encroaching on the northern frontier in India. When they occupied some territory in Bengal, the Governor-General decided that the limit of British patience had been reached and an army was assembled to penetrate the mountain passes and punish them. The 1st Battalion of the Regiment was included in this force. (Colonel Jasper Nicolls, of the XIVth, was serving as the British Army's Quartermaster-General at the time, and he was also involved – being appointed to the command of an Indian Infantry Brigade.)

This expedition was the first serious campaign into the great hill ranges which surround the north of India and the operations were not easy. In the event, the Regiment's contribution to their overall success was small, as it was withdrawn to garrison Dinapur before the campaign had really got under way. But the men of the XIVth gained valuable experience, which was soon to stand them in good stead.

In Central India the situation was steadily deteriorating when the XIVth was ordered to Agra. Trouble was brewing because the Rajah of Hatrass would pay his taxes only when the British Raj threatened military action. Hatrass itself was a fortified town, encircled by a great moat of uncertain depth, and when the Regiment arrived to bring the recalcitrant Rajah to heel Colonel Watson decided that ultimately an assault crossing would have to be made over this moat. In the dead of night a nameless volunteer achieved ephemeral fame by dangling a stone into the moat to determine its depth. When the Rajah persisted in his refusal to pay up, his capital was bombarded and the assault crossing took place. The town quickly surrendered to Watson, but a full scale attack had to be made on the Rajah's castle. In the fracas the Rajah himself escaped, but the Government installed a more amenable ruler and peace was restored to Agra.

The next serious challenge to British authority demanding the service of the Regiment occurred in 1825, when Baldeo Singh, the Rajah of Bhurtpore, died. This particular potentate had maintained friendly relations with the British Government. But the nephew, who promptly and illegally stepped into the dead man's shoes, was decidedly anti-British. Moreover the self-appointed Rajah, Doorjun Sal, was popular and his anti-British feelings were shared by a large number of his subjects.

When Doorjun Sal raised an army, and Bhurtpore's neighbouring states showed signs of following his lead in asserting independence, the Government decided it was time to deal with him. So a punitive force was assembled, and in December, 1825, this force arrived at the rebel capital. Two hundred miles south of Delhi, in what is now Rajputana, Bhurtpore was a large, densely populated and fortified town with a garrison numerically superior to the British force. Under the overall command of General Viscount Combermere, the latter had been organized into two divisions: one commanded by the redoubtable Jasper Nicolls (a Major-General by this time) and the second by Major-General Thomas Reynell. The XIVth, 900 strong and commanded by Major Matt Everard, was in Reynell's division.

When it became clear that the self-appointed rajah had no intention of giving way, Combermere ordered his artillery to smash a breach in the eight miles of fortifications. But when five weeks of continuous bombardment had made little impression on them Combermere ordered an assault engineering operation, and under cover of a bombardment sappers stacked two great mines of gunpowder under the corners of one bastion of the fortifications. When these were detonated on 18 January, 1826, the explosion blew two great gaps in the walls of the fort. Headed by its grenadier company, the XIVth raced through these gaps into the town.

Many of the grenadiers had been at Waterloo with the 3rd Battalion and their élan was undoubtedly a major factor in the success of the assault. But their casualties were high, the Regiment losing 136 killed and wounded in the course of the day. One subaltern, wounded when the mine went up, had his leg amputated on the spot but lived for another thirty years. By the later afternoon it was all over. Bhurtpore surrendered, a new rajah was installed with an officer of the XIVth, Colonel McCombe, as Governor to guide him, and General Reynell published a Divisional Order praising the Regiment's performance: 'It is impossible to convey half of what I feel in appreciation of the gallant conduct of the XIVth Regiment, which led the principal storming column. It has impressed my mind with stronger notions of what a British regiment is capable when led by such men as Major Everard.'

Finally, the story would not be complete without a brief mention of the fact that nearly half a million pounds was awarded in prize money for this operation – one of the last on land for which the government signified their gratitude in a mode which could be appreciated by the soldiers.

Jasper Nicoll's brilliant deceptive tactic is also noteworthy. The rebel Jats of Bhurtpore were terrified of the British grenades, and when they saw men advancing carrying grenades with lighted fuses more often than not they would break and run. Because these bombs were prone to explode prematurely, British soldiers also had a healthy respect for grenades with burning fuses. So for this operation

Jasper Nicolls ordered dummy grenades with burning fuses to be carried – relying wholly on the moral effect on the Jats. Last of all, reference must be made to the Bhurtpore gongs – great bronze gongs, three feet in diameter, which were seized by the XIVth as spoils of the siege. One of these gongs was mounted and stood outside the guardroom at the entrance to all barracks occupied by the XIVth during the remainder of their service in India. With other trophies the gongs were lost when the troopship carrying half of the Regiment to the West Indies was sunk off Guadaloupe on Christmas Day, 1837.

CHAPTER 7

The Crimea

THE XIVth's tour of duty in India came to an end in January, 1831. When the men with unexpired service had been posted to other units, a regimental cadre sailed from Calcutta to recruit and refit in the south of England. The move was effected at a time of great change in the British army. The 'seven years' service was abolished, new and better pensions were introduced, the last of the pikes – which had been carried by sergeants on ceremonial occasions – made their final appearance, and officers were shorn of much of the rich lace that had embellished their ceremonial dress since Regency days. The shako had already been substituted for the cocked hat, a coatee had replaced the old long-skirted coat, and trousers had taken the place of breeches and gaiters. Sartorial elegance was the order of the day; civilians had begun to wear tall hats, tail-coats and trousers, and the army was loath to fall behind in fashion. In the XIVth the changes brought the replacement of silver lace by gold, and the wearing of white coats by the band instead of buff coats faced with red.

After five years in England the Regiment sailed for the West Indies. These islands were still the most dreaded garrison stations of the colonies, as an anecdote by Colonel John Dwyer will illustrate. Dwyer in 1837 was a subaltern commanding a detached company of the XIVth on one of the smaller islands in the Barbados. 'Yellow Jack' was raging and when his company had lost twenty-one men to it in the course of a single week Dwyer sent a letter to Regimental Headquarters on St Kitts, asking grimly what he should do when the rest of his company was dead. 'Move now' came the reply.

After the West Indies came four years in the more congenial climate and terrain of Canada. The Regiment arrived at Quebec on 28 May, 1845, and that same night an horrific fire gutted the centre of the town. Lieutenant Tom Hamilton and the fiancée who had come out to marry him in Quebec were caught in the theatre and perished. As most of the buildings in Quebec were of wood, the conflagration spread rapidly, and the men of the XIVth worked like trojans to check the blaze and help those who were made homeless. 'The men were quickly assembled, and never relaxed an instant in their exertions,' the garrison commander reported warmly to the Commander-in-Chief of British troops in Canada – adding, 'I wish it had fallen to my lot to report . . . an occasion more honourable and more congenial to their profession.' When this report ultimately reached the British Army's supremo in Whitehall, the Iron Duke himself, the garrison commander's qualification was ignored. 'I am grateful,' the Duke noted, 'at this fresh display of all good qualities of British officers and soldiers – cool deliberation and energy in saving the lives and property of her Majesty's subjects.' It was a well deserved encomium.

For the rest of the time in Canada regimental life appears to have been uneventful and focussed on mundane matters. Rations in 1843 permitted only two meals a day: breakfast at 8 am, when a pound of bread and a pint of coffee were served, and dinner at 1 pm of a pint of soup, three quarters of a pound of boiled beef with two pounds of vegetables. The bread was described by one old soldier as 'infamous'. He had seen it to stick fast when it was flung against a brick wall, he said. An insistence on spit and polish also gave rise to some contemporary comment. Being 'dirty' on guard-mounting parade was a heinous crime, and at one such parade the adjutant is reported to have said to one man 'You have dirty hands and a dirty face; you have dirty buttons and dirty lace. Right about face! Double! And be damned to you.'

In 1847 the XIVth returned to England where it was re-equipped with the new percussion musket then being issued to all infantry regiments. Other important changes in the army's organization and equipment were also taking place. Unlimited service enlistments gave way to ten years initial engagement, with pensions after twenty-one years service; a new shako, known as the 'Albert hat' was introduced, and many other alterations in dress were in train. Unfortunately none of the innovations could alter the fact that the British army had stagnated since Waterloo. As a weapon it was old and rusty and another war had to be fought before any radical improvements came about in techniques, equipment and tactics.

This time it was a war fought for no reason at all, except a vague fear of Russia, and an amiable compliance with the grandiose ambitions of a French emperor destined later to lose his throne. When hostilities opened the XIVth was due for foreign service, and the news was greeted with an excitement bordering on exhilaration. An announcement that the Regiment was not to be included in the expeditionary force therefore came as something of a disappointment. However, the Regiment performed a vital service, garrisoning the Malta base, during the three major battles of Alma, Inkerman and Balaclava, and so missed the extreme privations of the early stages of the Crimean campaign. The men concerned could regard themselves as fortunate. Inevitably their turn came in February, 1855, when the siege of Sebastopol had begun. For six weeks the XIVth loaded and unloaded stores in a sea of mud at the tiny base port of Balaclava. The stores were badly needed in the forward areas where men were dying like flies of exposure and fatigue. But there was not enough transport to carry them there, barely enough fodder to feed the wretched animals used to pull what transport there was, and the road up to the front had degenerated into a slushy track. Three hundred new pairs of ammunition boots were issued to men of the XIVth employed on fatigue duties. But the heavy clay sucked the soles away from the uppers, and until replacements were forthcoming the wretched individuals concerned worked in their stocking feet on frozen ground at sub-zero temperatures. Poorly clad and wholly unaccustomed to the near arctic conditions of the Crimean winter, frostbite and pneumonia took their toll during the first two months. With spring the weather improved, and the privations due to cold were replaced by the horrors of cholera.

In March, the Regiment moved to the front, to join General Sir Richard England's 3rd Division and take its turn in the trenches. The machinery of war was nowhere near so deadly and destructive as it is today, but it would be wrong to think that life in the trenches of the Crimea was merely a matter of dodging cannon balls. While the artillery of the day was not radically different from that used in the Peninsular War, the guns were not toys. The solid iron ball was a deadly and erratic missile, and when it hit its effect was terribly destructive. Shells were also coming into vogue and a 'whistling dick' from a Russian battery could annihilate a couple of men without leaving any trace of them. Other novelties like the 'bouquet', consisting of a number of grenades enclosed in a larger one, contributed further to danger. But the Regiment acquired a reputation for luck or good management in the trenches. And after a long casualty-free spell this brought them the sobriquet 'The old Bombproofs'.

The siege of Sebastopol dragged on until September when the British assaulted the Russian fortification known as the Redan, while the French attacked the Malakoff. Although the Regiment was supposedly in reserve during this action it did contribute just over 200 men to Lt-Col Waddy's storming party – just over half the party's strength. In the action three of these men were killed and four others wounded. Although the British failed to take their objective the French capture of the Malakoff tower eventually resulted in the Russians evacuating Sebastopol. This brought the fighting to an end, and in March 1857 a peace treaty was signed.

Mercifully the Regiment suffered comparatively few battle casualties – 16 killed, and 46 wounded. (The number of those who succumbed to sickness and disease was considerably more but is not recorded.) The Crimea was evacuated in April, 1856, and the Regiment returned to Malta. Every man who had taken part in the campaign received Queen Victoria's Crimean medal together with a special silver medal struck by the Sultan of Turkey. In due course the Battle Honour 'Sebastopol' was inscribed on the Regimental Colours, the Commanding Officer, Colonel Maurice Barlow, got the CB; five officers, including Barlow, and one sergeant received the Sardinian Medal for valour; and three sergeants and three private soldiers were awarded French military medals.

So ended the last war to be fought in full dress uniform and the last in which the Colours were carried into battle. Just over a century later it is remembered principally by a cavalry charge, the heroism of Florence Nightingale, and the introduction of the Victoria Cross. Sixty years were to elapse before Britain was to become involved in another European war and in that time the remodelling of the British army was to change the form of the Regiment.

CHAPTER 8

The Maori Campaign

APART from the remote Crimea, no British soldier set foot on European soil for nearly a century after Waterloo. But there was much to do elsewhere. In 1857 the Indian Mutiny and sinister rumblings from other places where the Union Jack flew over countless trading-centres, factories and mission stations, led to another expansion of the British Army. So, for the second time in its history, the XIVth spawned another battalion. And it was this battalion which was to earn the Regiment's next laurels.

Raised in Ireland, most of the 2nd Battalion's new recruits came from around Liverpool and the Curragh. These men were young and tough – like their predecessors at Waterloo – and in December, 1858, when Colours were presented, the GOC Ireland said that he was 'well pleased' with their appearance and steadiness. By 1860 the battalion was judged to be fit and ready for foreign service, and in January of the following year it sailed for New Zealand, where British interests were being disrupted by a quarrel with some of the Maori chiefs.

The Maoris, an intelligent, athletic and high-spirited people, were a nation of fighting men skilled in savage warfare who had developed an extraordinary aptitude for building fortifications. From the white men who had come to New Zealand to trade, they had soon learned that if the Maori race was to escape extermination they would have to have firearms and to know how to use them. So native flax was traded for arms and ammunition, and rifle-pits and covered trenches were added to the defensive fortifications or *pas* they had already erected to check European encroachment on Maori territory. New Zealand was annexed by the British crown in 1840 and the Maori chiefs ceded their territories to Queen Victoria in return for the rights and privileges of British subjects. After that all went well for some years. But the continued influx of large numbers of British emigrants led to dissatisfaction, and this ultimately degenerated into open antagonism.

Fighting had already started when the 2nd Battalion arrived at Auckland, and the Regiment was flung straight into a campaign of bush warfare. Apart from the problems posed by the *pas* – which had been strategically placed and built with great ingenuity – the chief difficulty lay in transport and supply. Outside the European settlements there were virtually no roads or bridges and very few horses, mules or cattle for that matter. Consequently the men of the 2nd Battalion spent more time clearing and cutting tracks and building roads – mainly in the Waikato district of the North Island – than in actual fighting. (The Regiment's affiliation to New Zealand's Waikato Regiment stems from this period.) Once the problem of roads and transport had been overcome, there still remained the problem of dealing with the Maoris. And this was not easy. The stockades surrounding the *pas* consisted of tree-trunks laced together with strong vines known as supple-jacks. As a pre-requisite to an assault a gap had to be made in these stockades and when the attack went in the Maori garrison almost always contrived to escape. Again and again a *pa* was surrounded, but there was always a ravine or a water-course by which the Maoris slipped away; and when the British – often maddened by heavy losses – broke into the fortifications, it was to find no one there.

In March, 1861, peace was patched up with the Maoris and it seemed as if the war in New Zealand was over. So the 2nd Battalion settled down to a new and intensive programme of road building. The XIVth's fighting prowess had already come to official notice, and the zeal displayed by the XIVth in more mundane matters now brought another well-deserved encomium. In the event the peace was short-lived; hostilities were resumed in May, 1863, and the men of the 2nd Battalion sloped their new Enfield rifles and marched to the Queen's Redoubt – a fort which commanded an important river crossing. From there they went on to attack, capture and occupy a series of *pas* which the Maoris had set up between the Waikato and their newly established capital of Ngaruawahia. At one such *pa*, Rangariri – 'Angry Heavens' – the Maoris defended a square redoubt located inside a massive palisade behind a ditch eighteen feet deep and twelve feet wide. Rifle pits protected the flanks and covered all the likely approaches. Even to an untrained observer this particular *pa* was in a strong position, but just how strong it really was did not become clear until the attack went in. 'At the cessation of the cannonade, on the order to advance,' Captain Alexander Strange reported, 'our skirmishers and supports advanced under a heavy enemy fire, availing themselves of such cover as the ground afforded, until within about fifteen yards of the enemy works, and lay down, continuing the firing to keep down that of the enemy, who in considerable number occupied the ditch and parapet of the redoubt.'

The ditch was occupied after a good deal of hard fighting, and a mine was laid under the redoubt. When this was detonated another assault resulted in the *pa* being captured. Unfortunately the bill for the action was costly in relation to its importance. The commanding officer, Colonel Austen, was fatally wounded early in the battle, and one of the company commanders, Captain J. Phelps, was shot in the groin. Phelps, a trained surgeon, knew that he had not long to live and his last orders were that others who had a better chance of survival should have their wounds dressed first. This example of devotion to duty was typical of the sense of duty and military professionalism wholly in keeping with the traditions of the Regiment, which the young 2nd Battalion displayed throughout the campaign.

In one way or another the Maori wars dragged on until 1866, and the 2nd Battalion left New Zealand two years later. In a novel form of warfare – warfare in which the leather stock and pipeclay, red coats and shakos, gave way to blue jumpers and forage caps – the XIVth had again showed its mettle.

The reputation gained in New Zealand was consolidated next in India, where in 1877 the 2nd Battalion joined the army in which the 1st Battalion had been serving for the past nine years. During these nine years great changes had been taking place in the British army which necessarily affected the Regiment. The triumph of the Germans over the French in 1870, the incessant calls on the army for garrison duty in the Empire, the breakdown of the administrative

W.Y.R.—4

services in the Crimea, and an uneasy feeling that there was no reserve behind the regular army had ultimately roused the public conscience. In 1871 the purchase of commissions and promotions was abolished, bringing to an end the usatisfactory system which had been in vogue since the Regiment was raised in 1685. This step ensured that good officers lacking the necessary means would not be passed over by others less efficient who could afford to buy their promotion. Unfortunately the change did not speed up promotion, and officers were still grumbling about slow advancement even forty years later. In 1871 also came another change fundamentally affecting the rank and file. To stimulate recruiting and create reserves to draw on in time of war, the normal term of enlistment was reduced to seven years with the colours and five with the reserve. The eventual effect was to produce an army of youngish men in place of one half composed of old sweats. Many other reforms were also made about this time – barracks and feeding were improved, and there were small increases in pay. None of the reforms was received with much enthusiasm, and the greatest of all – making permanent the pairing of battalions and regiments, and officially abolishing the old numbers – raised a storm of protest. Under this new 'Cardwell' system the XIVth (Buckinghamshire) Prince of Wales's Own Regiment became The Prince of Wales's Own (West Yorkshire) Regiment, assigned to the 14th Regimental District based on York. With the brass '14s' on the men's shoulder straps went the time-honoured buff facings. White replaced buff, the letters 'w. YORK' replaced the old numbers, and the Prince of Wales's plume was substituted for the Royal Tiger badges worn on the soldiers' collars. In the same upheaval West Riding militia and country volunteers were affiliated to the Regiment and the outcry and bitterness expressed to these new fangled notions was analogous to that which followed the announcement of another series of reforms seventy five years later, when the proposed amalgamation of the XIVth and XVth Regiments was announced.

Other changes aroused less comment, because their importance was not fully appreciated. Probably the most significant was with the infantry's weapons. The Enfield muzzle-loading rifle was superceded by the Snider breech-loader, and then by the Martini-Henry.

The 2nd Battalion got their Martini-Henrys in 1874, fought both the Afghan and South African wars with them, and were very satisfied. By the end of the century there was smokeless powder, Enfield magazine rifles, new equipment, better clothing and more changes in dress.

In 1896, when the 2nd Battalion took part in an expedition in West Africa, scarlet uniform jackets were worn by British infantry on active service for the last time in more than two hundred and fifty years. King Prempeh of the Ashantis had resumed the barbaric practices which his predecessors had agreed to stop, and the expedition was sent to bring him to heel. To get to his capital, Kumasi, meant cutting a route through the bush. But Prempeh's habitual state of intoxication rendered him incapable of any resistance when the Battalion reached Kumasi, and so no fighting took place. In the event Prempeh was captured, dethroned and the Regiment took him back to the coast, where they re-embarked for England. It was a wholly successful albeit arduous expedition.

CHAPTER 9

Mainly South Africa

THE 1st Battalion moved to Gibraltar in 1895, where there was a brief reunion with the 2nd Battalion. The latter, on its way home from Aden, was diverted to Ashanti where trouble was brewing. From the Rock the 1st Battalion sailed for the Far East, serving in Hong Kong and Singapore and then moving back to India to stand guard behind the Khyber. When the Second Boer War erupted in October, 1899, the 1st Battalion remained in India, but the 2nd Battalion was among the first reinforcements to be sent out to South Africa. Under the command of Lieutenant-Colonel F. W. Kitchener – brother of Lord Kitchener of Khartoum – 27 officers and 936 rank and file embarked at Southampton on 20 October to disembark at Durban. Sir George White, with the bulk of the troops in South Africa, was then shut up in Ladysmith, and the 2nd Battalion was part of the force sent out to relieve him. Within a fortnight of their arrival the men of the 2nd Battalion were in action twenty miles south of Colenso. The Boers were occupying a range of hills covering the Tugela river, and the Battalion was ordered to seize Willow Hill, a prominent feature in their defence line. The objective was successfully occupied in a well conducted night attack, but at dawn the Battalion was treated to a demonstration of the tactical skill and marksmanship which made the Boer farmers such first class fighting men. When the brigade plan miscarried and the Boers brought a machine gun up to a position dominating Willow Hill, the Battalion's position became untenable, and Colonel Kitchener was ordered to pull back. In two days hard fighting 11 men were killed, 50 others wounded, and nothing was gained. Worse was to come, and this action was unhappily indicative of the fighting in the months ahead.

By November, 1899, it was clear that the Boers had settled down to starve out the garrison of Ladysmith, and the initial operation around Willow Hill had shown that relief of the besieged town was not going to be as easy a task as had been envisaged. Responsibility for the operation rested with General Sir Redvers Buller, and for the next three months the 2nd Battalion was continuously engaged in efforts to cross the Tugela river and dislodge the Boers blocking the approaches to Ladysmith. Throughout this period the Boers were holding a string of positions running for thirty miles, from Spion Kop on the west to Peter's Hill on the east, along the heights north of the river. The town of Colenso, twelve miles south of Ladysmith on the south bank of the Tugela and in the middle of the battle zone, became the focus of the fighting. The Tugela itself was about fifty yards across, with steep banks. There were no bridges but it could be crossed at several fords or 'drifts'.

On 15 December Buller ordered an attack on Colenso, and the 2nd Brigade – of which the 2nd Battalion formed a part – attacked with great élan. But the Boers were well dug in, and when it was seen that their marksmanship was taking a heavy toll the attack was called off. As a result, casualties were considerably less than they undoubtedly would have been if the battle had continued. Nevertheless the operation could only be counted a reverse. Coming as it did at the end of a week of British defeats, its only beneficial effect was to bring home to the British public the full realization of the situation in South Africa. The result was that Whitehall began to prosecute the war with greater vigour; this meant a flood of recruits and the eventual despatch of the 4th Militia Battalion to South Africa.

A second attempt to cross the Tugela was made in January, 1900. Heavy rain hampered the operation, and the 2nd Battalion had more trouble through the swollen torrent than with the Boers. By 22 January, however, Buller's troops had reached the low-lying hills, dominated by the massive Spion Kop and occupied by the Boers. Up to this phase of the operation the Boers had not shown any determined resistence. But Spion Kop was the key to their defence line and a tough fight was expected for this feature. Moreover Spion Kop was such a naturally strong position that it was concluded that only a night attack could succeed. In the event it was cold and drizzling with rain when the men of the 2nd Battalion stumbled up the narrow path to the Boer stronghold. In the dark the Boers appeared to have abandoned the hill and the West Yorkshiremen were deployed along what was thought to be the crest. But when morning dawned they found that the hill had neither been abandoned, nor were they occupying the summit. Devastating fire rained down from three sides. This went on all day – killing Captain Ryall, one of the company commanders, and five men, and wounding another officer and 42 men. For three days the Battalion grimly held its ground, and mercifully only one other man was wounded in this time. After the first day's shattering experience holes had been scooped in the rocky ground and stone sangars thrown up round them. But such a situation could clearly not continue, and Buller decided the game was lost – that it only remained for his troops to retire and try elsewhere. So the evacuation of Spion Kop was ordered, and by dawn on 27 January the 2nd Battalion was back across the Tugela – having marched all night. 'Men were exhausted, but coffee was made, and rum issued, and men marched cheerfully into Camp' the Regimental chronicler recorded.

A week later a new attack was undertaken – this time against the left of the new Boer front, where only the peak of Vaal Krantz blocked the road to the open country beyond. But when the Tugela had been crossed again and the Vaal Krantz hill captured, it was found that guns could not be brought up. Yet without artillery support a further advance would be too costly to risk. Meanwhile the 2nd Battalion had taken over a sector of the defences on Vaal Krantz Hill, and had lost no time in building sangars and digging in. It was hard work, but its benefits were reflected in the casualty figures. During the hours of daylight Vaal Krantz was shelled and sniped at without respite, yet only one officer and four men were slightly wounded in the twenty four hours of West Yorkshire occupation.

Orders to evacuate Vaal Krantz were received with some indignation; the morale of the troops was high and they could not appreciate the reason why the operation was considered to have failed. But the withdrawal was made, and the 2nd Battalion moved back once more to the south side of the Tugela.

The final attempt to break through to Ladysmith began on 18 February. This time the Boer left was the objective of the attack, and the operation opened with an action to secure the Monte Christo ridge five miles east of Colenso. Two battalions, the Queens on the right and the West Yorkshires on the left, advanced in open order across a bullet-swept plain. Bold action and the skilful use of cover made for success, and the crest was captured at the cost to the Regiment of 5 dead and 42 wounded. Six days of methodical fighting followed, and at the end of it only the Boers' northern buttress, Pieter's Hill, barred the way to Ladysmith. Its capture was not long delayed, the West Yorkshires taking a principal role, with Captain Conwyn Mansel-Jones setting a heroic example which won him the Victoria Cross.

A cavalry column entered Ladysmith soon afterwards, and the West Yorkshires were justly proud of their contribution to the operations which had made the town's relief possible. But this was not the end of the war. The Boers were not defeated yet, and for the 2nd Battalion much hard fighting lay ahead. Yet there was something of a respite about this time, during which Colonel Kitchener was promoted to command a brigade and Major Fry, Battalion second-in-command, took over. In April, 1900, a composite volunteer company joined the Battalion: H. D. Bousfield – who commanded the 1/5th Battalion in France and subsequently the Leeds 'Home Guards' – was one of the subalterns in this draft.

In February, 1901, the Regiment gained its second VC. Once the Boers in Natal and the Orange Free State had been dealt with, the British Army turned its attention to the Transvaal. From Ladysmith Buller's troops advanced northwards as part of a pincer manoeuvre designed to trap the Boer army commanded by General Botha. But as the trap closed Botha made a determined effort to break loose. During the night of 5 February, 1901, a powerful force of Boers crept up to the outpost line manned by the 2nd Battalion. Driving loose horses and cattle in front of them to stampede the outposts the Boers attacked in the pre-dawn dark and mist. But the West Yorkshires, who bore the brunt of the attack, stood firm; they had learned a lot about Boer tactics since Mill Hill, and the attack was repulsed with heavy Boer losses. During this action Sergeant William Traynor saw a wounded man who had been caught in the open and although the area was under heavy fire Traynor went out to bring him in. Within minutes Traynor himself had been hit, but with the help of Lance-Corporal Lintott the wounded man was dragged to safety. He survived, while Traynor, who was in great pain, carried on until the attack was over. That evening the GOC, Major-General Sir Horace Smith-Dorrien issued a special order of the day: 'The conduct of the West Yorkshires ... was especially fine, and their heavy losses are deplored.'

Another less successful action, typical of the guerrilla fighting of the last six months of this war, involved the 4th Militia Battalion – which served for two years on the lines of communication. In Cape Colony a convoy of wagons, guarded by sixty Colonial Mounted Rifles and a hundred West Yorkshire Militiamen was attacked in February, 1902, by Boer commandos. The convoy was laagered between two sangared kopjes, but the sangars gave insufficient protection and were too far apart to afford mutual support. The escort fought hard but thirteen Militiamen were killed or wounded and the rest overwhelmed. The captured men were released within a week and in a sorry state they eventually got back to the 4th Battalion.

In May, 1902, the war came to an end. It had cost Britain 28,000 casualties of which 345 were suffered by the Regiment. In the course of the fighting two VC's, one CB, eight DSO's and fifteen DSM's were awarded to officers and men of the 2nd Battalion and the 4th Militia also won two DSO's and three DCM's – a most creditable record. It is also relevant to add that Queen Victoria died in 1901, before the war ended, and one of her activities is still remembered as a motherly gesture. Each British soldier in South Africa received a box of chocolate bars in a special tin box with a portrait of the Queen and a message of goodwill on the lid. One such gift is now displayed in the Regimental museum – still intact, although the tissue lining of the box has yellowed with age and the chocolate is spoiled.

Like their fellow infantrymen the West Yorkshires emerged from 'the last of the Gentlemen's wars' thoroughly convinced of the value of realistic tactics and good shooting. After the war, army training included 'field firing' practice, in which officers were required to specify the target, the range, and the type of fire to be delivered, instead of merely regulating the discharge. From now on Physical Training started to replace some of the rigid barrack square work and route marches in full kit were introduced. So, although nobody realized as much at the time, the experience of South Africa did much to prepare the British soldier for the ordeals of the First World War.*

The 2nd Battalion returned to England in 1903 and spent the next eight years in Ireland and England. In 1911 it went to Malta, and the following year provided a contingent for the international force put into Scutari when the Turks were defeated during the Balkan War. Meanwhile, in 1908, the 1st Battalion had seen a spell of active service on the North-West Frontier of India against the Mohmands.

* At that date (1902), an infantryman's pay was a shilling a day with 3d 'messing allowance'. Free quarters were provided for wives and children in barracks, but there was no 'marriage allowance' in cash and no family rations in Britain. A married soldier was issued with personal rations separately – one pound of bread, three-quarters of a pound of meat a day, with an allowance of 3d a day to balance the extra rations received by single men.

CHAPTER 10

The Auxiliary Forces

THE Militia had its origins in the old compulsory levies of Saxon days but it was not until William Pitt initiated a reform of Britain's military structure that a proper citizen army was established. Under the terms of the Militia Act of 1757 every able-bodied man between the ages of 18 and 50 – with the exception of peers, members of parliament, clergymen, constables, apprentices, seamen, and any others who could 'establish proper grounds for exception' – became liable for service in the militia. Uniforms were free, pay was issued for exercise days, and adjutants and sergeants were seconded from the regular army. The new Act laid down a quota of three officers to every eighty men, and with a three-year rota of service it was hoped to achieve a sizeable reserve of manpower, reasonably well trained, to be called in on an emergency. One difficulty was a great shortage of officers, and in some counties several months passed before the necessary appointments were made. Not surprisingly the biggest county contribution to the militia quota came from Yorkshire – to the number of 2,360 or just over 7 per cent of the national total. Yorkshire also had the distinction of providing the first regiment to be raised under the Act. This was Colonel Thornton's Regiment – later the 3rd West York Light Infantry, and commissions in it were granted by the King on 27 January, 1759.

The 2nd West Riding Militia, formed by Sir George Saville as The York Regiment in 1759, was officially embodied in the following September. With a strength of some 650 all ranks it served as a garrison and anti-invasion force in Nottinghamshire until the end of the Seven Years War. It was dispersed in 1762, but called out again in 1778 during the American Revolution. In 1773 its title was altered to the 2nd (Northern) West Riding Militia; thirty years later this was amended to Yorkshire (West Riding) 2nd Regiment –

a designation which was retained until the era of the Crimean War. For five years, until its demobilization in York in 1783 the 2nd West Riding Militia was kept busy on internal security duties – mainly in the south of England. On occasions these duties were more than mere garrison routine. In June, 1780, for instance, the Gordon riots – precipitated by a sectarian act of parliament – had to be put down by force of arms, with considerable civilian casualties.

In 1793 the militia was called out yet again after the outbreak of the long war with revolutionary France, and it remained in active being for twenty three years, with only a brief break during the Peace of Amiens in 1802 and 1803. During this long period it was employed all over the United Kingdom on anti-invasion duties, guarding prisoners of war, and a host of other tasks normally carried out by regular soldiers.

According to the terms of the Militia Act, unless an individual could claim exception, service in the militia was supposed to be compulsory. However, no one was required to serve in person if he could provide a suitable substitute. As the price of these rapidly increased, many men who might otherwise have enlisted as regular soldiers found it more profitable and a good deal less arduous to become militia substitutes, and this had an adverse effect on recruiting for the regular Army. This was finally realized by the authorities and as the danger of invasion receded the men of the regular militia were lured into the regular Army by the offer of bounties. This made the regular militia ineffective for other purposes and most of its duties were eventually taken over by various supplementary and local forces.

The militia was stood down in 1816, and virtually ceased to exist until 1852 when the probability of war with Russia again made it necessary to consider the raising of auxiliary troops to replace the regular troops who would be sent abroad to fight the war. By that time, of course, practically nothing remained of the old and reasonably efficient force of Napoleonic days except for a few elderly gentlemen who were still listed as holding militia commissions. When it came to reforming the 2nd West York Light Infantry, the only equipment that remained of the old York (W.R.) 2nd Regiment was a few ancient flintlocks which had been dumped in York thirty-eight years previously.

In the emergency of 1853, however, the new force was quickly raised and trained. The 4th West Yorkshire Militia came into being that year, followed by the 2nd West York Light Infantry in 1854. The existence of the 4th, raised originally as the Leeds Regiment, was short, being disembodied in 1856. But the 2nd West York Light Infantry volunteered for active service and in June, 1855, it embarked for Gibraltar to relieve a regular battalion for service in the Crimea. The Battalion remained there until May, 1856, when it was relieved by the 92nd Highlanders and returned to England to be demobilized in York. In acknowledgment of the Regiment's service, the honour 'Mediterranean' was authorized to be borne on the 2nd West York Light Infantry Colours – which were eventually laid up in York Minster. As at the turn of the century, the period of 'non-existence' was short. In 1857 the outbreak of the Indian Mutiny resulted in the 2nd being called out on service once more. But it never left York, was quickly demobilized, and thereafter slipped back into its usual peace-time existence, only coming together at annual camps.

In 1881, as part of the Cardwell reforms, the 2nd West York Light Infantry and the 4th West York Militia became the 3rd and 4th Militia Battalions of the Regiment. And soon after the outbreak of the South African War, both battalions were embodied. As has been recorded in the previous chapter the 4th Battalion served for two years in South Africa, while the 3rd Battalion went to Malta. Additionally, both battalions sent off drafts to the 2nd Battalion. Both battalions stood down in 1902, every man receiving ten shillings travel money, and thirteen shillings and sixpence with which to buy himself a suit, a cap and a muffler. When these were put on and the men left barracks it was the end of the Militia era.

Volunteers first appeared at the end of the eighteenth century when hundreds of units were raised all over the country for home defence. These had almost all been disbanded by 1815, and it is certain that the ones raised in Yorkshire had no connection with those raised almost half a century later.

In 1859 war with France seemed certain. And, as the steamship, the railway and the telegraph had made it possible to mobilize large armies swiftly and secretly, the threat of a French invasion was very real. With the bulk of the British Army in India there were few regular troops to meet it.

Because there had been so many popular revolts elsewhere in Europe, the Government was reluctant to raise volunteer units. But as there seemed to be no alternative, Palmerston, the Prime Minister, eventually gave his consent. Initially no money was forthcoming from government sources, but public enthusiasm and patriotism was such that hundreds of units quickly came into being – the cost of arms, uniforms and other essentials being met by public subscription or by the individual members of the various corps. The movement was encouraged by the Prince Consort, who sensibly advised the adoption of a simplified system of drill and manoeuvre which would be within the capacity of part-time soldiers. In his opinion, he said, the new force should be trained so as to 'act efficiently as an auxiliary to the Regular Army and Militia; the only character to which it should aspire'. This in fact was its final role.

The Volunteer movement got off to a good start in Yorkshire's West Riding. Four battalions were soon raised and training began. The new units wore a simple inconspicuous uniform and were armed with the short muzzle-loading Lee Enfield rifle. As in the Home Guard of the Second World War, the intention was that any invader would be faced at every turn by hundreds of riflemen fighting over familiar terrain.

In 1872 the 1st West Riding Volunteer Battalion merged with the 1st York Battalion, and in 1883 – after the Cardwell reforms – the 1st (York), 2nd (Bradford) and 7th (Leeds) West Riding of York Volunteers became the 1st, 2nd and 3rd Volunteer Battalions of the Prince of Wales's Own (West Yorkshire) Regiment. The international scene was quiet at the time and no one could see any use for the Volunteers, but they soldiered doggedly on at week-ends, in the face of a good deal of ridicule and lampooning by such elegant journals as *Punch*. At the end of the century their turn came. The situation in South Africa was grave, and to boost the available manpower, Volunteers were encouraged to opt for active service with their regular battalions. Many served with distinction with the 2nd Battalion.

Following the South African War numerous committees were set up to consider how best to improve the weakness in the British military system that had been revealed. As a result of their deliberations, various recommendations were made which resulted in sweeping changes in 1908. Known as the Haldane reforms after the Secretary of State who introduced them, they were based on the assumption that Britain needed two military forces. The Regular Army would provide overseas garrisons and expeditionary forces in the event of war, while a new Territorial Force – organized primarily on a county basis was to be formed for home defence. The latter was to be based on the existing Volunteers, but was to be a balanced force of all arms with a proper permanent staff of regular soldiers to assist it.

Simultaneously the militia ceased to exist as a field force and became an organization for training men to reinforce the regular army during a major war. The old 3rd and 4th Militia Battalions thus became the 3rd and 4th Special Reserve Battalions, and the 1st, 2nd and 3rd Volunteer Battalions became the 5th, 6th, 7th and 8th Battalions of the Territorial Army. The 8th Battalion came into being as an offshoot of the 7th Battalion – the successor to the old 3rd Volunteer Battalion. Both the 7th and 8th Battalions were based on Leeds, and both held the additional title The Leeds Rifles. The 5th and 6th Battalions – both bearing the Regimental title – were based at York and Bradford respectively. Thus at the outbreak of the First World War in 1914 the order of battle was:

1st Battalion (Regular)
2nd Battalion (Regular)
3rd Special Reserve Battalion
4th Special Reserve Battalion
5th Battalion (Territorial Army)

6th Battalion (Territorial Army)
7th (Leeds Rifles) Battalion (Territorial Army)
8th (Leeds Rifles) Battalion (Territorial Army)
Regimental Depot.

CHAPTER 11

The First World War

THE military operations in which the Regiment was involved between 1914 and 1918 were of a type and scale previously undreamed of. Yet, although they varied in detail from front to front there was a terrible sameness about them all. In Europe, the main theatre, both sides went to ground early, and thereafter the war became a series of vast siege operations of which the British Army had had little experience since Sebastopol, sixty years earlier.

The trench lines were often close together and the armies in constant contact. Defence was superior to attack, there were no flanks to be turned, there was no way to manoeuvre and no way to hide massive troop movements. Yet both sides feverishly tried to punch a hole in the other's front by sending waves of infantry into infernos of artillery and machine gun fire. Shot down by the hundreds of thousands, most attacks gained little more than a few muddy shell-holes. Even on quiet days many men were killed and wounded as a result of casual shelling, sniping and raids; in the frequent offensives launched by both sides, battalions regularly suffered casualties halving and quartering their strength. Thus, in a book of this size it is impossible to do much more than to indicate briefly the part played by the various battalions against this general background.

In August, 1914, the 1st Battalion was at Lichfield as part of the 18th Brigade of the Sixth Division. Seven weeks after the declaration of war it was in action on the Aisne. It saw a good deal of hard fighting during the German drive to capture the Channel ports and out of the 1,364 casualties sustained by the 18th Brigade in the fighting on the Aisne more than half were of the West Yorkshires. Eleven days after the 1st Battalion landed in France only five officers and 250 other ranks were left of the original twenty-seven officers and 959 other ranks. Those who remained were suffering the first discomforts of trench warfare. 'Some of the men's feet were very sore from wet and mud . . . eighteen inches of water and six inches of mud . . . parapets keep falling in . . . men spend their time standing in one or two feet of water.' Over the next four years, the trenches were to become a way of life; in the mud and stench men managed to live and even found a few comforts.

The 2nd Battalion, recalled from Malta when the war started, arrived at Le Havre on 5 November, 1914. Within a week it was in the line near Ploegsteert Wood ('Plug Street') and went into action west of Neuve-Chapelle on 18 December. Like those of the 1st Battalion, the men of the 2nd Battalion were generally mature and well trained. But the swift change from the Mediterranean to a winter in the trenches took its toll, and although battle casualties were initially light there was a steady loss of men due to the new ailments of 'trench fever', 'trench feet' and frostbite. The 2nd Battalion was in the line on Christmas Day, 1914, and shared the brief and unofficial truce observed on many parts of the front. Early in the New Year sickness was rife and the low-lying trenches became so waterlogged that they were abandoned in favour of sand-bagged breastworks. These were more comfortable but they were also more vulnerable, and there was a sudden rise in the casualty rate.

The 2nd Battalion played its full part in many of the great battles of the war. It fought at Neuve-Chapelle in 1915 and did well; in 1916 it was on the Somme and on 1 July of that year lost sixteen officers and 490 other ranks. During a lull in that battle an observer noted that the No Man's Land between the trenches from which the 2nd Battalion had assaulted and the German lines was 'littered with motionless forms'. These were the bodies of West Yorkshire-men, lying in serried ranks right up to the parapet of the German trenches. July, 1917, found the reinforced survivors in the mud and chaos of the Ypres salient, where casualties were equally heavy.

The 5th, 6th, 7th and 8th (Territorial) Battalions formed the 1st West Riding Brigade of the 1st West Riding (Territorial) Division, and on 3 August, 1914, they were at their annual camp. Although the Territorial Army existed primarily for Home Defence, all four battalions were invited to volunteer for unlimited service; this caused much comment – the general tenor being that those who had bothered to prepare themselves in peacetime were now being asked to extend their liability. But almost every man did volunteer, and on 10 August the 1st West Riding Brigade was concentrated at Selby, where the Battalions were renumbered 1/5th, 1/6th, 1/7th and 1/8th West Yorkshires.

The Brigade spent the next eight months in hard training, and by the time the troops crossed the Channel in April, 1915, Battalion diaries were recording that the men were 'longing for service overseas'. By the end of the month the Brigade had not seen any Germans, but they had suffered casualties from shelling, and were beginning to settle down to the uncomfortable and dangerous routine of trench warfare. Although this brigade of West Yorkshiremen saw no large-scale action during its first twelve months in France, by April, 1916, all four battalions had suffered considerable casualties from small arms, shelling, gas, and disease. Manpower was also beginning to become a serious problem, and the old sweats who had formed the original units were sadly missed. Those who trickled back after recovering from their wounds were especially warmly welcomed, and the new men who came as reinforcements were gradually assimilated. Thus, throughout the war this exclusively West Riding brigade managed to maintain the Regimental standards and retain its peculiarly territorial character.

When the Somme battles began in September, 1916, the 1/6th, the 1/7th and 1/8th were in the thick of the fighting. And during the battle of Thiepval – 'Bloody Thiepval' as it was so often called – the same three gallant battalions held this mighty hinge of the German line. A year later they took heavy punishment at Passchendaele where the spongy, slimy state of the ground, churned up by constant shell fire produced appalling and unforgettable conditions. Finally, during the battles of the Marne in 1918, the 1/8th Battalion won a signal honour conferred by the French on only three British regiments. During July, under the energetic command of Lieutenant-Colonel Norman England, this battalion captured the Montagne de Bligny – a hill held by a numerically superior German force and swept by German machine gun fire. For taking this position, and holding it against desperate and determined counter-attacks, the

Regiment was awarded the Croix de Guerre which now rests in York Minster.

Each of the original first-line Territorial units spawned a second battalion early in the war. As the 2/5th, 2/6th, 2/7th and 2/8th Battalions these composed the 185th Brigade of the Sixty-Second Division, and at Cambrai in November, 1917, they took part in the great tank action which is associated with that battle. The 1st Battalion, as part of the 18th Brigade, was also involved. After a record advance spearheaded by the 1st Battalion, the attack ran into stiff opposition in the course of which the newly fledged 185th Brigade proved that its units were of the same stock, skill, and resolute material as its predecessors of the old XIVth. But this was the only major battle in which the second line Territorial battalions were destined to take part. Manpower and reinforcement problems dictated the disbandment of the 2/6th and 2/8th in January, 1918, of the 2/7th in June of that year, and of the 2/5th in August. With the exception of the 2/5th most of the men of the disbanded battalions were re-absorbed by the parent first-line units of the Regiment.

Early in the war the Regiment also raised eight Service Battalions. The 9th, which came into being in August 1914, took part in the ill-fated Dardanelles operation. After reaping laurels in Gallipoli it returned to take part in the battle of Thiepval – the same action in which Corporal George Sanders of the 1/7th Battalion won the VC. The 10th Service Battalion went to France in July, 1915, and in the battle of Fricourt exactly one year later German machine guns mowed down twenty-two of its officers and 750 other ranks. The 11th Service Battalion followed the 10th to France in August, 1915, and after the mud and blood of the Somme and Ypres was moved to the assistance of the Italians. After marching ninety miles in six days, the Battalion took up positions south of the Piave. On this front in October, 1918, during the triumphant advance to Vittorio Veneto, it captured over one thousand demoralized Austrian prisoners and four guns.

Like the second-line Territorial battalions, the existence of the 12th Service Battalion was concluded before the war ended. Moving to France in September, 1915, this battalion represented the Regiment at Loos. Having only just arrived in France, its men had no experience of trench warfare and had never been under fire. When the battle started they were also tired, hungry, thirsty and soaking wet. Yet they put up a performance that would have been a credit to an experienced, battle-tried unit.

The 15th, 16th and 18th were three splendid battalions of Leeds and Bradford 'Pals' forming the 93rd Brigade, which assaulted a German strongpoint in the battle of the Somme in 1916. Shortly after this action all three battalions were decimated in an attack on Serre, and Sergeant Yates of the 18th Battalion wrote, 'Someone pushed me in a bed of ferns. There were flowers among the ferns, and my last thought was a dull wonder there could still be flowers in the world.' In February, 1918, the reconstituted 16th and 18th Battalions of Bradford Pals were disbanded, but the 15th Battalion of Leeds men gained a new lease of life in November, 1917, when it absorbed the remnants of yet another Service battalion from Leeds. This unit, known as the 17th (Bantam) Battalion because of the size of its first recruits, had already won fame in a brilliant attack at Longueval.

Then there were two 'pioneer' battalions – the 21st (Pioneers) and the 22nd (Labour) – which also saw overseas service. Both went to France in the middle of 1916 to work on the lines of communication and the fact that the 21st Battalion alone suffered 182 casualties is an indication that their work as pioneers was no sinecure.

Finally this account would not be complete without a brief mention of those Regimental units which served the fighting battalions – the Depot, and the 3rd and 4th Special Reserve Battalions – through whose ranks many thousands of men passed on their way to the front and which, after the war, were responsible for the demobilization of men from all battalions of the Regiment. Two Home Service Garrison Battalions also enjoyed a brief existence – as did a 5th Reserve Battalion; a 3/5th, a 3/6th, a 3/7th and 3/8th offshoots of the original Territorial Force; the 19th, 26th Provisional, 51st and 52nd (Graduated) Battalions, 53rd Young Soldiers, the 6th and 9th Training Reserve Battalions.

The Regiment came out of the First World War with credit. It was only to be expected that the Regular battalions should have behaved well, but in the event the untried Territorial and Service battalions proved especially worthy successors to the men of Corunna, Waterloo and a score of other battles. Thousands of young men, who had never in their wildest dreams thought of being soldiers, passed through its ranks; sixty-six Battle Honours and four VC's were won, and thousands of men died in winning them. But at the end of it all the survivors were proud to be able to call themselves 'West Yorkshires'.

CHAPTER 12

The Second World War

THE end of the conflict which had been hopefully labelled 'the war to end all wars' did not bring an end to the British Army's tasks. Under the terms of the Peace Treaty there was an occupation zone around Cologne to be garrisoned. There was trouble in Egypt, Palestine and Iraq; while India was soon to reap the bitter aftermath of war in the shape of a conflict with Afghanistan, a series of arduous frontier expeditions, and a period of prolonged internal unrest. With the exception of the war with Afghanistan the Regiment was concerned with all these affairs.

Shortly after the signing of the Armistice the 1st and 8th Battalions marched into Germany; later the 1/5th and 1/6th Battalions also served with the Rhine army. In 1921 the Cardwell system came into operation once more and the 2nd Battalion went to India, where it took part in operations in Iraq – returning to England via the Sudan in 1930. Six years later the 2nd Battalion's services were needed to help to suppress an Arab rebellion in Palestine. In a difficult terrain, this task proved to be hard and incessant toil with little to show at the end of it.

Meantime, after a three-year spell in Germany, the 1st Battalion embarked on what was to be a long tour of foreign service. A spell in Jamaica and Bermuda – no longer the pestilential plague spots of a century earlier – was followed by a move to less sybaritic surroundings in Egypt. In 1934 the flies and mosquitos of the Canal Zone were forsaken for seven years in the Kiplingesque environment of India. Flag marches, security duties, a shattering earthquake in Baluchistan – where the battalion helped to salvage what was left of the devastated city of Quetta – and endless training were the highlights of this period. It ended with dramatic suddenness two years after the declaration of war against Germany, with orders for the Battalion to embark for Rangoon.

For the Territorial units of the Regiment the late 'twenties and early 'thirties were lean years, in which the 5th, 6th, 7th and 8th Battalions struggled against a chronic shortage of money and equipment and a contemptuous public attitude to maintain a creditable standard of strength and efficiency. Between 1937 and 1939 however, the renewed threat of war brought great changes. The 5th Battalion was again ordered to form a 2/5th Battalion; this, like its parent unit, continued to serve as infantry. The 6th Battalion was issued with searchlights and became the 49th AA Regiment RE; the 7th (Leeds Rifles) took to armour, to become the 45th Battalion Royal Tank Regiment; and the men of the 8th Battalion (Leeds Rifles) were initiated into the mystique of anti-aircraft gunners to become the 96th HAA Regiment RA. On paper the two militia battalions had continued as Regimental units since 1918 but theirs was a tenuous existence, and they were ultimately disbanded in 1953.

During the 'phoney' war of 1939 and 1940, the 1/5th and 2/5th Battalions were in England, training hard and aspiring to action. For the 1/5th this aspiration never materialized; in April, 1940, the Battalion was issued with leather jerkins, given a crash course in mountain warfare and told it would be going to Norway. The Battalion had actually embarked when a series of orders, counter-orders and confusion eventually brought a cancellation of the original instructions, and a month later it was sent to garrison Iceland – an island dubbed 'Hell's half-acre' by the American troops who joined them there two years later. Amid the scarcely veiled hostility of the Icelanders, the 1/5th soldiered on until April, 1942, when it returned to spend the rest of the war in the U.K.

To the 2/5th fell the lot of upholding the Territorial traditions of the Regiment. In April, 1940, it was sent to France for second-line duties at St Nazaire. At that time there seemed little immediate prospect of fighting, but the unloading of ships interspersed with guard duties was more attractive than endless training in England. In the event the end of the 'phoney' war heralded a wholly unforeseen change in circumstances, and the 2/5th became involved in bitter fighting around Dunkirk. On their return the remnants were reformed as a infantry unit. But in July, 1942, the 2/5th first became the 113th Regiment, Royal Armoured Corps and secondly a Tank Transporter unit. As such its brief existence lasted until September, 1943, when it reverted to an infantry role, and was renamed the 14th Battalion West Yorkshire Regiment.

The outbreak of the war had no immediate impact on the role of the 2nd Battalion. The Arab insurgents were still proving troublesome, and the situation was exacerbated by illegal Jewish immigration into Palestine. But a move to the Sudan was ordered in November, and when Italy entered the war in June, 1940, the Battalion went into action at Gallabat on the Ethiopian frontier. By February, 1941, however, the tide had turned against the Italians, and the West Yorkshiremen chased those who had attacked Gallabat up into the Wahni mountains. Then came the brilliant operation which gave Britain her first victory of the war and smashed Italian military power in East Africa.

Eritrea was the key to the Italian empire in East Africa, and the Keren gorge is the gateway to Eritrea. In their determination to hold this defile the Italians had deployed their strategic reserve in wellnigh impregnable positions dominating every approach to it. Over a period of six weeks, desperate fighting for the cliffs at the entrance to the gorge failed to do more than gain an insecure foothold on the lower slopes. So an all-out frontal attack was decided upon, and under cover of a formidable artillery barrage this attack was launched on the morning of 15 March, 1941. The West Yorkshire objective was the great bastion of Mount Dologorodoc, 1,500 feet above the valley and crowned by a fort; this, the north countrymen assaulted from the west. Surprised and stunned by the ferocity of the assault the Italians fought back desperately. But at dawn on 16 March the West Yorkshiremen were in the fort and had taken some 400 dazed Italians and Eritrean prisoners. For the ten more days that the battle lasted, exposed on three sides to Italian artillery and mortars, the men of the 2nd Battalion remained an immovable and decisive wedge in the hostile line. They ran out of ammunition and food, which had to be dropped on the rocks from old Vincent and Wellesley aircraft. They were bitterly counter-attacked eight times

– by fresh troops, Eritreans, Alpini, Bersaglieri, Savoia Grenadiers – and every attack was driven back. Finally the Italians broke, Keren fell, and the road to Asmara, the capital of Eritrea, was open. Keren now ranks as a Battle Honour, and no embroidery of the facts is necessary in asserting that it was a great feat of arms. And it was recognized and welcomed as such at the time by an English speaking world sorely in need of encouragement.

After Keren the Italians were finished in East Africa. At Ad Teclesan a half-hearted attempt was made to slow the British advance to Asmara, but the garrison of Savoia Grenadiers had no stomach for fighting and they surrendered en masse to the 2nd Battalion. When the Asmara and the post of Massawa had been secured it was time for the Battalion to seek laurels elsewhere. Thus June, 1941, found the 2nd Battalion in camp at Quassasin, twenty-five miles west of the Suez Canal, preparing for a campaign in the Western Desert. In August it moved up to build the reserve positions which later became famous as the Alamein Line, but it was not long there. At 3 am on the morning of 22 August urgent orders arrived for an immediate move to the Mediterranean coast, and the rumour was that the West Yorkshires would be air-lifted to China.

The real destination and the reason for the move soon became clear however. The 5th Indian Division was on its way to Iraq, to take joint action with the Russians against pro-German elements in Persia. But, by the time the long dusty trek across Palestine ended at the oil fields of Kirkuk, a new Shah had been installed and a new treaty with Persia made the journey unnecessary. So the 2nd Battalion retraced its steps towards the Western Desert, only to be diverted to Cyprus. With the Germans in possession of Crete and with their aircraft operating from Rhodes, Aphrodite's island was a potential battlefield. In the event the months slipped by without incident and during April the Battalion returned to its old camp at Quassasin.

In December, 1941, the Japanese launched simultaneous attacks on Pearl Harbour, Hong Kong and Malaya. Singapore, symbol of British strength in the Far East, fell in February, 1942, and a

Japanese army was advancing into Burma. In an effort to stave off disaster, an infantry and an armoured brigade was rushed to Rangoon from India.

The 1st Battalion disembarked at Rangoon in the middle of an air raid on 2 February and entrained for a region close to Pegu, in support of the hard-pressed 17th Division. The latter, disorganized and battleworn, was falling back before a superior Japanese force, and barely a month after their arrival in Burma the West Yorkshires were in action. Four officers and twelve other ranks were killed, two officers and twenty-one other ranks were wounded, and seventy-three others reported missing in the battle of Pegu. (Of these seventy-three, thirty-four were captured by the Japanese but only twenty-three of them survived the war.)

During the battle great numbers of Japanese troops had been driving past to the north of Pegu and, when they reached the Rangoon–Prome road, the fate of Rangoon was sealed. The evacuation of the town was ordered and the long weary retreat back to India began. In an effort to stop the troops getting back the Japanese had erected a succession of road-blocks past which the retreating columns had to force their way. A series of difficult and dangerous delaying actions had to be fought before contact was finally broken and what was left of the Burma army finally got away across the Chindwin. With the 1st Battalion in the rearguard the 800 mile retreat came to an end in May at Tamu in the Khabaw Valley. Since leaving Rangoon the Battalion had suffered 173 casualties. But it had gained a great reputation for dogged endurance and unfailing courage, under the famous general who was to become the Regiment's next Colonel.*

Back in the Middle East the British faced General Rommel and a German and Italian Army in Cyrenaica. By April, 1942, the battle line had been stabilized at Gazala, when the 2nd Battalion took up position at Halfaya Pass ('Hellfire' Pass) near the Egyptian frontier on the coastal road from Mersa Matruh to Sollum. The pass was a hundred miles behind the firing line but it was a very suitable place

---

* Field-Marshal Sir William Slim (later Viscount Slim) succeeded Field-Marshal Sir Cyril Deverell as Colonel of the West Yorkshire Regiment in 1947.

to become acclimatized to desert life. Nothing grew on the red cliffs of the Sollum escarpment, water was rationed to three-quarters of a gallon per man per day, and the tide of battle had left the whole locality strewn with booby traps.

In the middle of May the West Yorkshires were ordered to move forward to defend part of the Tobruk perimeter. And so, after 5,000 miles of travel without seeing a single enemy, the 2nd Battalion advanced to participate in what is now considered to be one of the most disastrous campaigns of World War II.

Rommel's offensive opened at the end of the month, and on 2 June the German armour was reported to have been caught in the Knights-bridge 'Cauldron' on the end of the Gazala mine marsh – without oil, petrol, water and rations. Orders for a 'liquidation' operation were issued and in the early hours of 5 June the West Yorkshires moved forward to the Bir Hakeim track. An epic of confusion fol-lowed. The tanks preceding the Battalion ran into very heavy fire from well dug-in anti-tank guns, which inflicted heavy losses and forced them to pull out. The infantry, left without armoured pro-tection, was then counter-attacked in the open by the far from im-mobile German armour and the Yorkshiremen were overrun. Only with great difficulty were the survivors extracted from the area and reconstituted as a fighting unit in the El Adem 'Box', which on 1 July was ordered back to the Nile Delta. Behind it the reformed 2nd Battalion left a small mobile column, 'Langcol', which was to have an adventurous career roving the desert and harrassing the enemy.

One other action fought by the 2nd Battalion at the end of August deserves mention, not least because it came at the turning point of the war. As expected, Rommel attempted to exploit his success by battering through the Alamein defences and when his thrust was blocked at Alam Halfa the Axis forces in Lybia had lost their last chance of victory in North Africa. In this, the First Battle of Alamein, the West Yorkshires – deployed on the bare and rocky Ruweisat Ridge – defeated a diversionary attack. But when the battle was over and they were relieved, the prolonged period in action had reduced the 2nd Battalion to a total of 300.

Reformed, reorganized and refreshed by a sojourn close to the flesh-pots of Cairo, the 2nd Battalion left Egypt in September, 1942, for garrison duty in Iraq. From there in May, 1943, it moved to India and in June of that year to the Burma front. Six months later the Battalion was in the Arakan, and the first Japanese blood was drawn at Kanyindan near Maungdaw. When Maungdaw was occupied the West Yorkshires pushed south against increasing resistence. In January, 1944, however, the 2nd Battalion was ordered to cross the famous Ngakyedauk Pass – better known by those who served with the 14th Army as the 'Okeydoke' Pass.

At the other side of the Pass the Battalion reached the 7th Division administrative base at Sinzweya just as the Japanese launched a furious counter-offensive. General Tanahashi's objective was nothing less than an invasion of India from Arakan, and the defensive 'box', hastily created by West Yorkshiremen and Gurkhas, was soon surrounded by hordes of 'Banzai'-yelling Japanese. The divisional commander, Major-General Messervy – who had known the Battalion well in Eritrea – had managed to fight his way to the box before his headquarters was overrun, reputedly in his pyjamas. The battle raged round Sinzweya from 6 to 22 February. Until the siege was broken supplies had to be dropped by parachute and sleep had to be snatched at the bottom of slit trenches with the rats. But the gallant 'Battle of the Admin Box' upset the Japanese plans and their invasion of India had to be postponed. In his despatch on the Arakan operations the Corps Commander, General Christison, wrote of the Yorkshires 'Never has any Regiment counter-attacked so successfully and so often as in that battle. It is rare in history that one Regiment can be said to have turned the scale of a whole campaign.'

Having failed to blast a way to India through the Arakan the Japanese now launched a great new offensive on the central Burma front. One hundred thousand crack Imperial troops were concentrated for a drive through the British base of Imphal in Manipur in an operation described by the Japanese commander, General Mutaguchi, as a March on Delhi. The main axis of advance lay along the Tiddim road, where the 'Black Cats' of the 17th India Division were deployed. These troops, who had been fighting the Japanese for more than two years, included the 1st Battalion.

As the Japanese began to converge on Tiddim in March, 1944, the 'Black Cats' began a fighting withdrawal, during which the 1st Battalion distinguished itself at 'Vital Corner', near the village of Tonzang. This action, in which General Mutaguchi's 33rd Division was given a bloody nose, enabled the rest of the Division to break through a series of roadblocks the Japanese had established along the divisional line of march. In a deluge of rain, what remained of the 1st Battalion eventually got to a concentration area about forty miles from Imphal town, and on 5 April occupied the 'Cat Fish' defensive box on the edge of the Imphal plain.

Two days later, in the unusual and somewhat macabre setting of a siege, an officer of the 1st Battalion established contact with the 2nd Battalion. The latter had flown to Dimapur from the Arakan with the rest of the 5th India Division, and been ferried down to Imphal. There they were deployed north-west of the town.

In the hard fighting which continued over the next few months both the 1st and the 2nd Battalion were heavily engaged. For some weeks, until Kohima was relieved by concentrated attacks in which the 2nd Battalion gained further laurels, the Japanese had disrupted the preparations for a British offensive in Burma. On 22 June, 1944, however – the 259th Anniversary of the formation of the Regiment – the siege of Imphal was lifted, and General Slim realized that the time had arrived to strike back. Thus at the beginning of July, the 'Black Cat' Division started to retrace its steps towards Tiddim, and on 10 July 'A' company of the 1st Battalion attacked a Japanese position at Ningthoukhong. The attack was successful but the appalling casualties consequent on a Japanese counter-attack spelled the end, for the time being, of the 1st Battalion as a first-line fighting unit. In a little over six months the Battalion had suffered 344 casualties in battle alone – including ten officers and ninety-eight other ranks killed. So the Battalion's next move was back to India to recoup and refit.

When the 1st Battalion left the battlefield the 2nd Battalion took its place, as the 5th Indian Division passed through the 17th to continue the advance on Tiddim. The monsoon was starting, malaria and stomach troubles were rife; and there was still plenty of fight in

Mutaguchi's disease-ridden survivors. But the operations had to go on, and in some of the most difficult terrain in the world the 2nd Battalion fought an action about every three miles along the 200-mile stretch of road. Men travelled light, relying on supplies dropped from aircraft, as they negotiated a series of steep ridges whose ascent, it was calculated, would have taken them twice as high as Everest. As the advance progressed and the monsoon weather deteriorated, however, the problems of getting reinforcements forward increased, and the strength of the 2nd Battalion steadily dwindled. By the time the 5th Division had climbed the 'Chocolate Staircase' to reach Tiddim in October, 1944, the Battalion had been reduced to the strength of a single company. Reconstituted as 'X' Company, the West Yorkshimen were placed under command of the 4th Royal West Kents, with whom they served until December. Finally, in December, 1944, 'X' Company returned to Kohima as the cadre of a new 2nd Battalion.

It was a new 1st Battalion which returned to the fray in January, 1945. Many of those who had seen service up to and including the rigours of Imphal and the Tiddim road had been repatriated or posted elsewhere. New drafts had brought new faces; one large batch of reinforcements had come from the disbanded 11th Battalion of the Regiment which had been stationed in the Falkland Islands. After staging at Imphal the Battalion set out for Meiktila, a focus of road and railway routes to Mandalay vital to the Japanese. Action came after crossing the Irawaddy, when the 1st Battalion was ordered to attack and clear the eastern half of Meiktila town. As an earlier attack by a Gurkha battalion supported by a squadron of tanks had failed to dislodge the Japanese defenders it was clear that this was going to be a tough job. And so it proved. In four hours of savage fighting the Regiment gained its second Victoria Cross of the campaign, but the objective could be pronounced secure when every single Japanese soldier had been accounted for. After the battle an assessment of Japanese casualties put the number at 235. In the course of the action nine West Yorkshiremen were killed and sixty-four wounded.

If the Japanese had not expected an attack on Meiktila, they were not slow to react to its capture. But as they began to concentrate for a counter-stroke, the 5th Indian Division was flown in to Meiktila. Serving with the 9th Brigade of this division was the 2nd Battalion, which – like the 1st – had risen from the metaphorical ashes of Tiddim. When the 1st Battalion attacked the village of Kyigon on 25 March, the men of the 2nd Battalion watched the action from the positions they had taken up round the nearby airfield. Subsequently both battalions were engaged in the mopping up operations round Meiktila.

From Meiktila the 1st Battalion moved in the 17th Division's advance towards Rangoon. Rangoon was 300 miles away, but the Battalion was in action no more than fifteen miles from Meiktila. A suicide squad had been left at almost every village on the route, and the liquidation of such squads invariably brought a sharp fight. Snipers were troublesome also, and sadly all these actions brought their toll of casualties.

In April, the 5th Division leap-frogged through the 'Black Cats' to push on and capture Toungoo. The race for Rangoon was now on, and as the pace increased Japanese resistance crumbled away. The only doubt now was how many Japanese troops scattered about Southern Burma would escape east across the Sittang.

The first of May saw the 17th Division – with the 1st Battalion in the van – in possession of Pegu, and on 2 May Rangoon fell to para-troops and to others who had landed from the sea virtually un-opposed. After that the work remaining to be done in Burma was merely 'mopping up', if the liberation of Malaya and of the Dutch East Indies still had to be accomplished. The advance on Pegu had scattered the Japanese, many of whom were now making their way east towards the Moulmein area from where they hoped to retire into Malaya. To intercept and round up as many of these fugitives as possible the 1st Battalion moved towards Moulmein to take post first at Mudon and then at Mergui – areas of derelict plantations whose overgrown state made patrolling very difficult. Large numbers of Japanese surrendered at both places, 123 Japanese officers handing over their swords at Mudon alone. Meantime the 2nd Battalion had moved into the difficult country east of Pegu. Patrolling, blocking the tracks and hunting down Japanese in this region led to some quite sharp encounters, and there was a sense of relief and new purpose when the 2nd Battalion went to Mingaladon to prepare for the proposed invasion of Malaya. Mercifully, the Japanese surrender on 15 August, 1945, took most of the danger out of the invasion, and 'Operation Zipper' encountered no hazards. And it was a fitting finale to the Regiment's war service in the Far East that the 2nd Battalion should be the first British Regiment to land in Singapore.

Finally, to conclude a long but necessarily abridged chapter of events which are relatively fresh in the chronicler's mind, those Regimental units which had an ephemeral war-time existence are worthy of mention. The 11th Service Battalion, formed at York in July, 1940, moved to Cornwall in October and to the Isle of Wight in July, 1941, returning to Bridlington in Yorkshire in February, 1942. In June that year it embarked for an unknown destination with a strength of thirty-two officers and 929 other ranks, as part of 'Task Force 122'. In the event, the unknown destination turned out to be the Falkland Islands where the Battalion remained until January, 1944. Many of those who served with this battalion subsequently saw active service, and the Battalion itself was eventually amalgamated with the 5th Battalion and a battalion of the Lancashire Fusiliers to form a training unit.

What happened to the 6th, 7th and 8th Battalions has been outlined already. Only the course of the 6th Battalion needs expansion. In 1940 Searchlight Regiments RE were taken over by the Royal Artillery, and four years later the old 6th Battalion became the 49th Garrison Regiment RA. As such it served in Antwerp during the V-bomb attacks. Subsequently it became the 601 Regiment RA (The West Yorkshire Regiment), which was reformed and renamed after the war the 584th Mobile HRA Regiment (West Yorks) RA (TA).

The 9th (Overseas Defence) Battalion was formed at Ripon in November, 1939. Most of its members were reservists of between thirty-five and fifty years of age, many of whom had served in World War 1. Moving to France in February, 1940, the 9th was, in the words of the Colonel of the Regiment who saw it off, 'the first Battalion of the Regiment to go out to France'. In the confused fighting around Dunkirk it is known to have acquitted itself well, although precisely what happened has disappeared into the limbo with the records which were lost at the time. Dunkirk spelled the end of the 9th Battalion, which was disbanded in June, 1940.

A 10th (Home Defence) Battalion, formed in York in November, 1939, by the amalgamation of untrained and young-soldier companies, was so oversubscribed by September, 1940, that it was split into the 1/10th and 2/10th Battalions. But the existence of the 2/10th was short-lived and it became the 13th (Home Defence) Battalion of the Regiment three months later. Re-designated the 30th Battalion in 1941 it garrisoned in the Scilly Islands until December, 1942, when it was disbanded. The 1/10th parent unit, which reverted to its original title 10th Battalion, was amalgamated with the 6th Battalion in September, 1942.

The 12th Battalion, which originated from the 50th Battalion – formed in the summer of 1940 from No. 5 Holding Battalion – was employed on the defences of the East Coast. When it was disbanded in August, 1941, a large proportion of its members were drafted to the 59th Battalion Reconnaissance Corps.

As mentioned previously, the 13th Battalion, formed in September, 1940, could be considered an offshoot of the 1/10th Battalion. After serving in the Scillies it was moved to Cornwall and disbanded in December, 1942.

Finally there was the 70th (Young Soldiers) Battalion. Formed at Leeds in September, 1940, this battalion guarded airfields and other important zones and vital points until April 1942. A change in role then ushered in a period of intensive training and the battalion was looking forward to active service overseas. But the prime purpose of this battalion had been to serve as a reinforcement training unit, and when the authorities decided it had outlived this purpose it was disbanded in September, 1943.

CHAPTER 13

The Final Years

SOUTH-EAST Asia was in a turmoil when Japan capitulated in August, 1945. Nationalist elements in Indo-China and Indonesia were striving to assert the independence of their countries before the pre-war colonial masters could re-impose their authority; in Malaya the Malayan People's Anti-Japanese Army was laying the foundation for communist subversion.

The Regiment was concerned with restoring order in both Indonesia and Malaya. Two days after VJ Day the Indonesian nationalists proclaimed an Indonesian Republic, and mobs in Sourabaya attacked the British and Indian soldiers who had gone there to rescue internees and round up the Japanese garrison. Reinforcements were rushed across to Sourabaya from Singapore, and the 2nd Battalion was among them. Two extremely hazardous and unpleasant months of security operations followed, and then when law and order had been restored in Java the 2nd Battalion had to start a similar task in Malaya. Ostensibly this was accomplished by August, 1946, although the Communist fanatics were, in fact, deep in the jungle preparing for the campaign they were to launch in 1948.

During 1947 the 2nd Battalion concentrated in Penang, and in May the following year it sailed from there to Liverpool, en route for amalgamation with the 1st Battalion in Austria. When the amalgamation took place at Klagenfurt on 24 August, 1948, 146 years had elapsed since the original 2nd Battalion was raised in Belfast.

Following this reorganization, reducing the fighting element of the Regiment to a single battalion, a brief interlude of peacetime soldiering ensued. But 1952 found the West Yorkshires in the Suez Canal Zone, where the state of sullen inertia which preceded an Egyptian revolution could hardly be considered peace. Before the final agony of the 1956 Suez operation, however, what was termed a Malayan 'Emergency' claimed the Regiment's service. The Communist insurrection which had been simmering since 1945 finally erupted in 1948, and although the crisis point was passed, the terrorist campaign was far from being defeated. With Battalion headquarters in Ipoh, West Yorkshire companies were deployed in Perak and Pahang. In this campaign company commanders controlled areas of responsibility sometimes as big as the West Riding, and executed operations aimed at isolating the 'Charlie Tares' and starving them out. Every now and then the companies would concentrate for a battalion or brigade operation to 'squeeze' a selected area. When agents obtained information concerning terrorist trails and lairs, patrols would disappear into the jungle – sometimes for weeks on end – to lay ambushes and disrupt the Communist links. Suspected terrorists' camping spots would be bombarded by the Battalions' mortars, to keep the terrorists on the move and sap their morale.

Slowly the Communist casualties grew – though not without cost to the Regiment. By the time the 1st West Yorkshires left Malaya to return to Northern Ireland the Communists had come to recognize that the bottom had been knocked out of their 'Liberation' movement.

After Malaya came an uneventful tour of duty in Northern Ireland, before the dramatic and controversial Anglo-French expedition against Egypt in 1956. When the 1st Battalion disembarked in Port Said, Operation 'Musketeer' was at a depressing stage with the West Yorkshiremen facing the humiliating task of clearing up the mess created by the initial assault. Their role was the usual one associated with internal security duties; law and order had to be maintained – this included the enforcement of a dusk to dawn curfew – and there were considerable quantities of arms at large which had to be found and collected. The town was divided into areas of responsibility and to maintain effective control of the land also meant controlling the surrounding water. The Battalion had been allotted the waterfront and to cope with this problem boats were requisitioned and a private West Yorkshire 'navy' raised to patrol the harbour.

The Suez Canal operation has been dubbed a political fiasco, but a military success. And the West Yorkshires, who were the last British troops to leave, would certainly contest criticism of their part in the affair. Any expression of regret would almost certainly be confined to the tragic affair which came at the end of the operation, when a promising young officer of the Regiment was abducted and subsequently found to have been murdered.

After Suez it soon became sadly and abundantly clear that the days of the West Yorkshire Regiment as an entity were numbered. The Defence White Paper of April, 1957, announced that National Service was to be discontinued and the size of the Army reduced to something near its recruiting potential. This meant the disappearance of a number of Infantry regiments; the decision as to which were to go being taken inevitably – although not necessarily very fairly – on ability to recruit. Regiments were not to be disbanded but were to amalgamate in pairs, surplus officers, warrant officers and sergeants being 'axed'. For the West Yorkshires the situation might have been much worse. Since their formation as the XIVth and XVth of Foot the two Regiments had been friends; and since the turn of the century they had been neighbours. Nevertheless the blow was a shattering one.

One of the first manifestations of the new order was the adoption of a Brigade cap badge; and the Yorkshire rose surmounted by a gilt crown with 'Yorkshire' in a scroll beneath the rose was adopted. The loss of the White Horse of Hanover after more than 200 years was severely felt; the cap badge had always been the symbol *par excellence* of the Regiment, and its disappearance forcibly brought home the finality of the proposed changes.

Inevitably the worst effects were felt in the Sergeant's Mess, the traditional stronghold of Regiment loyalties. Since the war many officers had served with other Regiments, or the Colonial Forces, or on the Staff, so that although their loyalty remained strong they were at least conditioned to change, whereas the bulk of the warrant officers and sergeants had spent their service in one or other of the battalions. When the conditions for premature discharge – the so-called 'golden bowlers' – were announced many members of all ranks applied to go. The steady reduction of the Infantry had reduced promotion prospects to vanishing point while industry was booming. And although these conditions might have been acceptable if the

Regiment had continued, the announcement of its demise was decisive. Over the months that followed there was a steady exodus of familiar figures from the Regiment to civilian life.

The sad but colourful amalgamation ceremony took place at Dover on 31 July, 1958, in the presence of HRH The Princess Royal. (Her Royal Highness, Colonel-in-Chief of the West Yorkshires since 1947, continued as Colonel-in-Chief of the new Regiment, The Prince of Wales's Own Regiment of Yorkshire.) When to the strains of 'Ca Ira' and 'The Yorkshire Lass' four representative guards – two from the old XIVth Foot and two from the XVth Foot – marched their Colours to the centre of the parade, a chapter closed. But in the West Riding there are still many thousands of men who wore the badge of the White Horse of Hanover, and most of their hearts and quite a lot of their memories remain with The West Yorkshire Regiment, the XIVth Foot.

Acknowledgments

In writing this brief history of The West Yorkshire Regiment I have been grateful for the help of Major H. A. V. Spencer, who provided me with much of the background material and checked facts. The originals of the illustrations and many other fascinating items of Regimental history are in his charge in the Regimental Museum at Imphal Barracks in York.

I also owe a great deal to my predecessors who have recorded the Regimental history in the following works:

Historical Records of the 14th Regiment, Captain H. O'Donnell
Historical Records of the Fourteenth or Buckinghamshire Regiment of Foot, Richard Cannon
The Records of the Third Battalion Prince of Wales's Own West Yorkshire Regiment or 'York Regiment', Colonel George Jackson Hay CB
A Peep over the Barleycorn, Charles James O'Mahoney ('Jack the Sniper')
The West Yorkshire Regiment in the War 1914–1918, Everard Wyrall
From Pyramid to Pagoda, Lieutenant-Colonel E. W. C. Sandes DSO, MC
A Short History of the Prince of Wales's Own Regiment of Yorkshire (XIV and XV Foot) 1685–1966, Major H. A. V. Spencer

APPENDIX A

The Regimental March

Ça Ira was composed by Bécourt in 1790 as dance music and known originally as the Carillon national. But its catchy tune was adopted by the French revolutionaries. Played by French regimental bands, the contemporary refrain reflected the French mood.

Ah! Ça ira, Ça ira, Ça ira; les aristocrats à la Lanterne,
Ah! Ça ira, Ça ira, Ça ira; malgré les mutins, tout réussira
was chanted by the crowds escorting victims to the guillotine.

After the battle of Famars Ça Ira was adopted as the XIVth Regimental Quickstep by express order of the Duke of York.

> And they play 'Ça Ira' yet
> In the old Fourteenth,
> In memory of the glorious day
> When they swept their foes away!
> In memory of the right begun
> When beneath the southern sun,
> To the Frenchman's tune they won
> The men of the Fourteenth.

ÇA IRA.

The March of the 14th P.W.O. Regiment.

The tune 'God Bless the Prince of Wales' also constituted a Regimental March.

APPENDIX B

The West Yorkshire Regiment
(*The Prince of Wales's Own*)

BADGES
The White Horse of Hanover with the motto *Nec Aspera Terrent*, and the Prince of Wales's Plume. Both are incorporated in the badges of the Prince of Wales's Own Regiment of Yorkshire, formed by the amalgamation of the old XIVtth and XVth Regiments. The 'Royal Tiger', superscribed 'India', worn on the buttons of the West Yorkshire Regiment is now borne on the Regimental Colour of the Prince of Wales's Own Regiment of Yorkshire.

REGIMENTAL ANNIVERSARY: IMPHAL DAY
22 June – commemorating the raising of the siege of Imphal 22 June, 1944, where both the 1st and 2nd Battalions fought against the Japanese. Also the day and month of the year that the Regiment was raised in 1685.

ALLIED REGIMENTS
The Waikato Regiment of New Zealand
The Royal Montreal Regiment of Canada
The 14th Battalion Australian Infantry (The Prahan Regiment)
The Falkland Islands Defence Force

APPENDIX C

Synopsis of Service

1685	Raised by Sir Edward Hales at Canterbury
1689–1692	*Beveridge*'s Regiment, served in Scotland and Flanders
1692–1713	*Tidcomb's Regiment (14th Foot)*, Flanders, England, Ireland
1713–1743	*Clayton's Regiment (14th Foot)*, Scotland, England, Gibraltar
1743	*Price's Regiment (14th Foot)*, Flanders, Scotland, England
1751	*The XIVth Regiment of Foot*
1752	Gibraltar
1759	England
1766	Nova Scotia
1771	America
1772	West Indies
1773–1777	America
1777–1780	England
1780–1781	Served as marines with the Royal Navy
1782–1809	*The XIVth Bedfordshire Regiment*
1782–1791	West Indies
1791	England
1793–1794	Flanders and France
1795–1803	West Indies
1805	Germany (1st Battalion)
1806	Ireland (1st Battalion)
	England and Ireland (Newly raised 2nd Battalion)
1807	England (1st Battalion)
1808	India (1st Battalion)
	Spain (2nd Battalion)
1809–1876	*The XIVth Buckinghamshire Regiment*
1809	Walcheren Expedition (2nd Battalion)
1810	Mauritius and India (1st Battalion)
1810–1814	Gibraltar, Malta, Sicily, Lampedusa (2nd Battalion)
1811	Java (1st Battalion)
1813	Borneo (1st Battalion)
1815	Nepal (1st Battalion)
1814–1816	Mediterranean area in the Napoleonic War (2nd Battalion) (2nd Battalion disbanded 1817)

1813–1816 3rd Battalion raised, served at Waterloo, disbanded
1815–1826 India
1831–1832 England, Ireland
1836–1841 West Indies
1841–1847 Canada
1847–1854 England, Ireland
1854 Malta
1855 Crimea
1856 Malta
1858 Greek Islands
2nd Battalion reformed in Ireland
1860–1864 West Indies (1st Battalion)
1860–1870 New Zealand, Australia (2nd Battalion)
1864–1868 England, Ireland, Malta (1st Battalion)
1868–1879 India (1st Battalion)
1870–1878 England (2nd Battalion)
1876–1881 *The XIVth (Buckinghamshire) Prince of Wales's Own Regiment*
1878 India, Aden (1st Battalion)
1878–1880 India (2nd Battalion)
1879–1895 England, Ireland (1st Battalion)
1881–1920 *The Prince of Wales's Own (West Yorkshire) Regiment*
1893 Burma (2nd Battalion)
1895 Gibraltar (1st Battalion)
1896–1899 Hong Kong, Singapore (1st Battalion)
1895 Gibraltar and the Ashanti War (2nd Battalion)
1896 England (2nd Battalion)
1899–1904 South Africa (2nd Battalion)
1899–1911 India (1st Battalion)
1904–1912 Ireland, England (2nd Battalion)
1911–1914 England (1st Battalion)
1912 Malta (2nd Battalion)
1913 Albania, Turkey (2nd Battalion)
1914–1918 *The First World War (38 Battalions)*
Service in Flanders, France, Gallipoli, Italy and Egypt
1919–1922 England, Germany (1st Battalion)
England, India (2nd Battalion)
1920–1958 *The West Yorkshire Regiment (The Prince of Wales's Own)*
1922 Iraq (Kurdistan) (2nd Battalion)
1926 N. Ireland (1st Battalion)
1924–1929 India (2nd Battalion)
1929–1930 Sudan, England (2nd Battalion)

1929–1931 West Indies (1st Battalion)
1931–1934 Egypt (1st Battalion)
1934–1946 India, Burma (1st Battalion)
1936 and
1938–1940 Palestine (2nd Battalion)
1939–1945 *The Second World War (8 Battalions)*
Service in India and Burma (1st Battalion); Sudan, Ethiopia, Egypt, Iraq, Cyprus, N. Africa, Burma (2nd Battalion); France and Belgium (2/5th and 9th Battalions); Falkland Islands (11th Battalion); U.K. (8th: 1/10th and 2nd/10th Battalions; 50th; 70th Battalions); (1/5th Battalion) Iceland
1946–1952 England, Austria (1st Battalion)
1945–1948 Malaya, Indonesia (2nd Battalion)
1948 England and finally to Austria for amalgamation with the 1st Battalion
1952–1953 England, Egypt (1st Battalion)
1953–1955 Malaya
1955–1956 N. Ireland
1956–1958 Suez operation in the Canal Zone of Egypt and then England
25 April, 1958 Amalgamation with The East Yorkshire Regiment